T0334588

WATER-BASED CHEMICALS AND TECHNOLOGY FOR DRILLING, COMPLETION, AND WORKOVER FLUIDS

WATER-BASED CHEMICALS AND TECHNOLOGY FOR DRILLING, COMPLETION, AND WORKOVER FLUIDS

by

JOHANNES KARL FINK
Montanuniversität Leoben, Austria

AMSTERDAM • BOSTON • HEIDELBERG • LONDON
NEW YORK • OXFORD • PARIS • SAN DIEGO
SAN FRANCISCO • SINGAPORE • SYDNEY • TOKYO
Gulf Professional Publishing is an imprint of Elsevier

Gulf Professional Publishing is an imprint of Elsevier
225 Wyman Street, Waltham, MA 02451, USA
The Boulevard, Langford Lane, Kidlington, Oxford, OX5 1GB, UK

Copyright © 2015 Elsevier Inc. All rights reserved.

No part of this publication may be reproduced or transmitted in any form or by any means, electronic or mechanical, including photocopying, recording, or any information storage and retrieval system, without permission in writing from the publisher. Details on how to seek permission, further information about the Publisher's permissions policies and our arrangements with organizations such as the Copyright Clearance Center and the Copyright Licensing Agency, can be found at our website: www.elsevier.com/permissions.

This book and the individual contributions contained in it are protected under copyright by the Publisher (other than as may be noted herein).

Notices
Knowledge and best practice in this field are constantly changing. As new research and experience broaden our understanding, changes in research methods, professional practices, or medical treatment may become necessary.

Practitioners and researchers must always rely on their own experience and knowledge in evaluating and using any information, methods, compounds, or experiments described herein. In using such information or methods they should be mindful of their own safety and the safety of others, including parties for whom they have a professional responsibility.

To the fullest extent of the law, neither the Publisher nor the authors, contributors, or editors, assume any liability for any injury and/or damage to persons or property as a matter of products liability, negligence or otherwise, or from any use or operation of any methods, products, instructions, or ideas contained in the material herein.

Library of Congress Cataloging-in-Publication Data
A catalog record for this book is available from the Library of Congress

British Library Cataloguing in Publication Data
A catalogue record for this book is available from the British Library

ISBN: 978-0-12-802505-5

For information on all Gulf Professional Publishing
visit our website at http://store.elsevier.com/

This book has been manufactured using Print On Demand technology. Each copy is produced to order and is limited to black ink. The online version of this book will show color figures where appropriate.

Working together
to grow libraries in
developing countries

www.elsevier.com • www.bookaid.org

CONTENTS

PREFACE

This manuscript focuses on water-based fluids with respect to chemical aspects. After a short introduction in the basic issues water-based fluids, the text focuses mainly on the organic chemistry of such fluids and special additives.

The nature of the individual additives and the justification why the individual additives are acting in the desired way are explained. The material presented here is a compilation from the literature, including patents. In addition, as environmental aspects are gaining increasing importance, this issue is also dealt carefully.

HOW TO USE THIS BOOK

There are four indices: an index of tradenames, an index of acronyms, an index of chemicals, and a general index.

In a chapter, if an acronym is occurring the first time, it is expanded to long form and to short form, e.g., acrylic acid (AA) and placed in the index. If it occurs afterwards, it is given in the short form only, i.e., AA. If the term occurs only once in a specific chapter, it is given exclusively in the long form.

In the chemical index, bold face page numbers refer to the sketches of structural formulas or to reaction equations.

ACKNOWLEDGMENTS

The continuous interest and the promotion by Professor Wolfgang Kern, the head of the department, is highly appreciated. I am indebted to our university librarians, Dr. Christian Hasenhüttl, Dr. Johann Delanoy, Franz Jurek, Margit Keshmiri, Dolores Knabl, Friedrich Scheer, Christian Slamenik, and Renate Tschabuschnig for support in literature acquisition. This book could not have been otherwise compiled. Thanks are given to Professor I. Lakatos, University of Miskolc, who directed my interest to this topic.

Last but not least, I want to thank the publisher for kind support, in particular Katie Hammon, and also Anusha Sambamoorthy and Kattie Washington.

J.K.F.

CHAPTER I

General Aspects

1. HISTORY

Some aspects of the history of drilling fluids can be found in monographs and in the Internet [1–8]. Also a compact time table of the history is available [9].

The earliest known oil wells were drilled in China in 347 before Christ. These wells had depths of up to some 240 m [10]. They were drilled by bits attached to bamboo poles. In 1845, the French engineer Pierre-Pascal Fauvelle (1797-1867) introduced a new method of well drilling [11–13]. He used hollow iron tubes to direct the water into the drill hole [14]. Drilling fluids were used around 1850 in the percussion drilling technique in order to suspend the cuttings.

In the early time, water-based fluids were used. Already the old Egyptians used water to remove the cuttings from their holes. In 1846, Robert Beart proposed that cuttings from drilling holes may be removed by water. Around 1890, clay was added to drilling fluids. At the same time, it has been proposed that a mixture of water and a plastic material—to be recycled—can be used to remove the cuttings. Such a mixture also should form an impermeable layer along the wall of the borehole [15].

Historically, water-based drilling fluids have been used to drill a majority of wells [16]. Their lower cost and better environment acceptance as compared to oil-based drilling fluids continue to make them the first option in drilling operations.

In addition, emulsion muds have been the systems of choice in drilling onshore, continental shelf, and deepwater wells to minimize the risk, maximize drilling performance, and reduce the costs [17]. However, because of environmental constraints, a high frequency of lost circulation, and the high unit cost of emulsion systems sometimes their benefits can not be fully emerge. Lost circulation materials and treatments have been reviewed [18].

Water-Based Chemicals and Technology for Drilling, Completion, and Workover Fluids
http://dx.doi.org/10.1016/B978-0-12-802505-5.00001-9

Copyright © 2015 Elsevier Inc.
All rights reserved.

2. FIELDS OF APPLICATION FOR WATER-BASED COMPOSITIONS

Water-based compositions are used in the petroleum industry for several purposes:

* drilling muds,
* fracturing fluids,
* cementing,
* filter cake removal,
* waterflooding,
* enhanced oil recovery,
* squeezing,
* transporting aids, and
* pipeline pigging.

For these operations, a basic fluid is needed and also in most cases certain additives are needed that are improving the performance of the basic fluid. For different purposes, for example, bacteria control or corrosion control, the same additive compounds can be used, at least with minor variations of the chemical structure. These general purpose additives are collected in special sections.

Guidelines for the selection of materials and the mechanism of corrosion in the petroleum industries have been summarized [19].

Tradenames appearing in the references are shown in Table 1.

Table 1 Tradenames in references

Tradename Description	Supplier
Biovis® Scleroglucan viscosifier [16]	Messina Chemicals, Inc.
Duovis® Xanthan gum [16]	M-I Swaco – Schlumberger
Flotrol™ Starch derivative [16]	M-I Swaco – Schlumberger
Glydril® Poly(glycol) (cloud point additive) [16]	M-I Swaco – Schlumberger
Poly-plus® HD Acrylic copolymer (shale encapsulation) [16]	M-I SWACO – Schlumberger
Polypac® UL Polyanionic cellulose [16]	M-I SWACO – Schlumberger
Polyplus® RD Acrylic copolymer (shale stabilizator) [16]	M-I SWACO – Schlumberger

REFERENCES

[1] Forbes RJ, O'Beirne DR. The technical development of the Royal Dutch/Shell: 1890–1940. Leiden: Brill Archive; 1957.

[2] Brantly J. History of oil well drilling. Houston: Book Division, Gulf Pub. Co; 1971. ISBN 9780872016347.

[3] Buja HO. Historische Entwicklung in der Bohrtechnik. In: Handbuch der Baugrunderkennung. Teubner: Vieweg; 2009. p. 17–25. ISBN 978-3-8348-0544-7. doi:10.1007/978-3-8348-9994-1_2.

[4] Buja HO. Handbuch der Tief-, Flach-, Geothermie- und Horizontalbohrtechnik Bohrtechnik in Grundlagen und Anwendung. Wiesbaden: Vieweg + Teubner; 2011. ISBN 9783834899439.

[5] Springer FP. Zur Geschichte der Tiefbohrtechnik aus der Perspektive von Lehr-und Fachbüchern. Res montanarum 2012;50:257–65.

[6] Apaleke AS, Al-Majed AA, Hossain ME. Drilling Fluid: State of the Art and Future Trend. The Woodlands, Texas: Society of Petroleum Engineers; 2012. ISBN 978-1-61399-181-7. doi:10.2118/149555-MS.

[7] Cass D. Canadian Petroleum History Bibliography. Calgary, Alberta: Petroleum History Society; 2013. URL: http://www.petroleumhistory.ca/history/phsBiblio2013-10.pdf.

[8] Clark MS. The history of the oil industry. Tech. Rep.; San Joaquin Valley Geology; Bakersfield, CA; 2014. URL: http://www.sjvgeology.org/history/.

[9] Russum S, Russum D. History of the World Petroleum Industry (Key Dates). Geo-Help, Inc.; 2012. URL: http://www.geohelp.net/world.html.

[10] Totten GE. Timeline of the History of Committee D02 on Petroleum Products and Lubricants and Key Moments in the History of the Petroleum and Related Industries. Incontext. West Conshohocken, PA: ASTM International; 2007. URL: http://www.astm.org/COMMIT/D02/to1899_index.html.

[11] Fauvelle PP. Sonda hidraulica y mejoras que la perfeccionan. ES Patent 306; 1861.

[12] Fauvelle M. On a new method of boring for artesian springs. J Frankl Inst 1846;42(6): 369–72.

[13] Kroker W. Fauvelle, Pierre-Pascal. In: Day L, McNeil I, editors. Biographical Dictionary of the History of Technology. London, New York: Routledge; 1996. p. 440. ISBN 9780415060424. URL: http://books.google.at/books?id= m8TsygLyfSMC&pg=PA440&lpg=PA440&dq=Fauvelle+patent&source=bl& ots=fXitCldcMo&sig=JA4BuR2jpB66hNHquvOlRJZi_dc&hl=de&sa=X&ei= QEcDU9fFJMmZtAap04DAAg&ved=0CF4Q6AEwBw#v=onepage&q=Fauvelle %20patent&f=false.

[14] Robinson L. Historical perspective and introduction. In: ASME Shale Shaker Committee, editor. Drilling Fluids Processing Handbook; Chap. 1. Amsterdam: Elsevier; 2005. ISBN 9780750677752.

[15] Chapman MN. Process of treating restored rubber, etc. CA Patent 35180; 1890.

[16] Patel AD, Stamatakis E. Low conductivity water based wellbore fluid. US Patent 8598095, assigned to M-I L.L.C. (Houston, TX); 2013. URL: http://www. freepatentsonline.com/8598095.html.

[17] Dye W, Daugereau K, Hansen N, Otto M, Shoults L, Leaper R, et al. New water based mud balances high-performance drilling and environmental compliance. SPE Drill Complet 2006;21(4). doi:10.2118/92367-PA.

[18] Alsaba M, Nygaard R, Hareland G, et al. Review of lost circulation materials and treatments with an updated classification. American Association of Drilling Engineers 2014; AADE-14-FTCE-25.

[19] Bahadori A. Corrosion and Materials Selection: A Guide for the Chemical and Petroleum Industries. Chichester, West Sussex, UK: Wiley; 2014. ISBN 1118869192.

CHAPTER II

Drilling Fluids

A drilling fluid or mud is a specially designed fluid that is circulated through a wellbore as the wellbore is being drilled to facilitate the drilling operation. The various functions of a drilling fluid include removing drill cuttings or solids from the wellbore, cooling and lubricating the drill bit, aiding in support of the drill pipe and drill bit, and providing a hydrostatic head to maintain the integrity of the wellbore walls and prevent well blowouts. Specific drilling fluid systems are selected to optimize a drilling operation in accordance with the characteristics of a particular geological formation [1].

For a drilling fluid to perform its functions, it must have certain desirable physical properties. The fluid must have a viscosity that is readily pumpable and easily circulated by pumping at pressures ordinarily employed in drilling operations, without undue pressure differentials.

The fluid must be sufficiently thixotropic to suspend the cuttings in the borehole when fluid circulation stops. The fluid must release cuttings from the suspension when agitating in the settling pits. It should preferably form a thin impervious filter cake on the borehole wall to prevent loss of liquid from the drilling fluid by filtration into the formations.

Such a filter cake effectively seals the borehole wall to inhibit any tendencies of sloughing, heaving, or cave-in of rock into the borehole. The composition of the fluid should also preferably be such that cuttings formed during drilling the borehole can be suspended, assimilated, or dissolved in the fluid without affecting physical properties of the drilling fluid.

Most drilling fluids used for drilling in the oil and gas industry are water-based muds (WBMs). Such muds typically comprise an aqueous base, either of fresh water or brine, and agents or additives for suspension, weight or density, oil-wetting, fluid loss or filtration control, and rheology control.

1. CLASSIFICATION OF MUDS

The classification of drilling muds is based on their fluid phase alkalinity, dispersion, and the type of chemicals used. The classification according to Lyons [2] is reproduced in Table 2.

Water-Based Chemicals and Technology for Drilling, Completion, and Workover Fluids
http://dx.doi.org/10.1016/B978-0-12-802505-5.00002-0

Copyright © 2015 Elsevier Inc.
All rights reserved.

Table 2 Classification of drilling muds

Class	Description
Freshwater muds[a]	pH from 7 to 9.5, include spud muds, bentonite-containing muds, phosphate-containing muds, organic thinned muds (red muds, lignite muds, lignosulfonate muds), organic colloid muds
Inhibited muds[a]	WBM that repress hydration of clays (lime muds, gypsum muds, seawater muds, saturated salt water muds)
Low-solids muds[b]	Contain less than 3-6% of solids. Most contain an organic polymer
Emulsions	Oil-in-water and water in oil (reversed phase, with more than 5% water)

[a]Dispersed systems.
[b]Nondispersed systems.

2. TYPES OF WATER-BASED DRILLING FLUIDS

The growing concern among government and environmental agencies over the environmental impact of drilling fluids has led to a significant increase in the industry's reliance on WBMs [3]. Actually, about 85% of all drilling fluid compositions used today are water-based systems. The types depend on the composition of the water phase, such as [3]:

- pH,
- ionic content,
- viscosity builders (i.e., clays, polymers, or a combination),
- filtration control agents,
- deflocculants,
- dispersants.

The mud composition selected for use often depends on the dissolved substances in the most economically available makeup water or on the soluble or dispersive materials in the formations to be drilled. Several mud types or systems are recognized and described in the literature as given in Table 3.

These types are explained below. In general, WBMs have water as the continuous phase. Water may contain several dissolved substances. These include alkalis, salts and surfactants, organic polymers in colloidal state, droplets of emulsified oil, and various insoluble substances, such as barite, clay, and cuttings in suspension.

Table 3 Mud types [3]
Mud type

Freshwater muds
Spud muds
Dispersed/deflocculated muds
Lime muds
Gypsum muds
Seawater muds
Salt water muds
Nondispersed polymer muds
Inhibitive potassium muds
Calcium-treated muds
Potassium-treated muds
Cationic muds
Mixed metal hydroxide muds
Low-solids muds

2.1 Freshwater Types

Freshwater fluids range from clear water having no additives to high-density drilling muds containing clays, barite, and various organic additives [3]. The composition of the mud is determined by the type of formation to be drilled. When a viscous fluid is required, clays or water soluble polymers are added. Fresh water is ideal for formulating stable drilling fluid compositions, as many mud additives are most effective in a system of low ionic strength. Inorganic and/or organic additives control the rheological behavior of the clays, particularly at elevated temperatures. Water-swellable and water-soluble polymers and/or clays may be used for filtration control. The pH of the mud is generally alkaline and, in fact, viscosity control agents like montmorillonite clays are more efficient at a pH more than 9. Sodium hydroxide is by far the most widely used alkalinity control agent. Freshwater muds can be weighted with insoluble agents to the desired density required to control formation pressures.

2.2 Seawater Types

Many offshore wells are drilled using a seawater system because of ready availability [3]. Generally seawater muds are formulated and maintained in the same way as freshwater muds. However, because of the presence of dissolved salts in seawater, more electrolyte stable additives are needed to achieve the desired flow and filtration properties.

2.3 Saturated Salt Water Types

In many drilling areas both onshore and offshore, salt beds or salt domes are penetrated [3]. Saturated salt muds are used to reduce the hole enlargement that would result from formation-salt dissolution through contact with an undersaturated liquid. In the United States, salt formations are primarily made up of sodium chloride. In other areas, e.g., northern Europe, salt may be composed of mixed salts, predominantly magnesium and potassium chlorides. It has become quite common to use high (20-23% NaCl) salt muds in wells being drilled in deep (>500 m water depth) water regions of the Gulf of Mexico. The reasons are twofold: stabilization of water-sensitive shales and inhibition of the formation of gas hydrates. The high salinity of salt water muds may require different clays and organic additives from those used in fresh or seawater muds. Salt water clays and organic polymers contribute to viscosity. The filtration properties are adjusted using starch or cellulosic polymers. The pH ranges from that of the makeup brine, which may be somewhat acidic, to 9-11 by the use of sodium hydroxide or lime.

2.4 Calcium-Treated Types

Fresh or seawater muds may be treated with gypsum or lime to alleviate drilling problems that may arise from drilling water-sensitive shale or clay-bearing formations [3]. Gypsum muds are generally maintained at a pH of 9-10, whereas lime muds (lime added) are in the 12-13 pH range. Calcium treated muds generally require more additives to control flow and filtration properties than those without gypsum or lime do.

2.5 Potassium-Treated Types

Generally potassium-treated systems combine one or more polymers and a potassium ion source, primarily potassium chloride, in order to prevent problems associated with drilling certain water-sensitive shales [3]. The flow and filtration properties may be quite different from those of the other water-base fluids. Potassium muds have been applied in most active drilling regions around the world. Environmental regulations in the US have limited the use of potassium muds in offshore drilling owing to the apparent toxicity of high potassium levels in the bioassay test required by discharge permits.

2.6 Low-solids Types

Fresh water, clay, and polymers for viscosity enhancement and filtration control makeup low-solids and so called nondispersed polymer muds [3].

Low-solids muds are maintained using minimal amounts of clay and require removal of all but modest quantities of drill solids. Low-solids muds can be weighted to high densities but are used primarily in the unweighted state. The main advantage of these systems is the high drilling rate that can be achieved because of the lower colloidal solids content. Polymers are used in these systems to provide the desired rheology, especially xanthan has proven to be an effective solids suspending agent. These low-solids muds are normally applied in hard formations where increasing the penetration rate can reduce drilling costs significantly and the tendency for solids buildup is minimal.

2.7 Emulsified Types

Classical additives for water-based oil–in–water emulsion fluids can be [4, 5]:
- emulsifiers;
- fluid loss additives;
- structure-viscosity-building substances;
- alkali reserves;
- agents for the inhibition of undesired water-exchange between drilled formations, e.g., water-swellable clays,
- wetting agents for better adhesion of the emulsified oil phase to solid surfaces; e.g., for improving the lubricating effect and for the improvement of the oleophilic seal of exposed rock formations;
- disinfectants for inhibiting bacterial attack on such emulsions.

WBMs, in diverse known formulations, perform satisfactorily in many applications. However, a particular problem arises when drilling through oil-bearing sand formations [6].

When the oil is thick and heavy it is difficult to recover. Conventional production well technology, which relies on crude oil flowing by gravity or pressure into production wells, does not work in bituminous oil sand formations. The recovery of heavy oil can be significantly enhanced using steam-assisted gravity drainage.

Using directional drilling methods, a horizontal production well is drilled through an oil sand formation. Then, a steam injection well with a perforated liner is drilled above and substantially parallel to the production well. Superheated steam is then injected into the oil sand formation. In this way, the oil is heated. This heating effect causes the oil to become less viscous. However, an effective removal of the cuttings from the mud is difficult and the ability to clean and reuse the mud is reduced [6].

These problems can be reduced by circulating the bitumen–laden mud through a mud cooler, which is an expensive method. Another promising solution for these problems is to use a polymer drilling fluid with some 0.3% of a nonionic surface active agent.

It has been found that the emulsification of the oil and bitumen in oil sand cuttings in a WBM may be effected by the addition of surfactants with hydophilic–lipophile balance (HLB) numbers equal to or greater than approximately 7.

The HLB of a surfactant is a measure of the degree to which it is hydrophilic or lipophilic. This term has been created by Griffin [7].

Effective surfactants are given in Table 4. These particular surfactants are manufactured by the Stepan Company, Inc., of Burlington, Ontario, Canada.

Nonyl phenol ethoxylate is shown in Figure 1.

Ester oils in water-based drilling fluids of the oil-in-water emulsion have an improved ecological acceptability [5]. These oils may be a 2-ethylhexylester of a C_8 to C_{14} fatty acid mixture, propylene glycol monooleate, or oleic acid isobutylester.

2.8 Inhibitive Types

The hydration of a water-sensitive formation during drilling can be prevented by using an inhibitive formulation [8]. The drilling fluid contains a high molecular weight nonionic poly(acrylamide) (PAM) with a molecular weight of 4–15 MD, a low molecular weight nonionic PAM with a molecular weight of 0.5–2 MD, long-chain alcohols, and polyanionic cellulose (PAC). An example of such a formulation is given in Table 5.

CLAY GRABBER™ is a high molecular weight PAM material, CLAY SYNC™ is a nonionic low molecular weight PAM, GEM™ is a long-chain alcohol, FILTER CHEK™ is a modified starch, and BARAZAN® is a viscosifier, all available from Halliburton, Houston, Texas.

Table 4 Nonionic surfactants [6]

Compound	HLB number	Trade name
Nonyl phenol ethoxylate	8.3	IGEPAL® CO–430
Octyl phenol ethoxylate	10	IGEPAL® CA–520

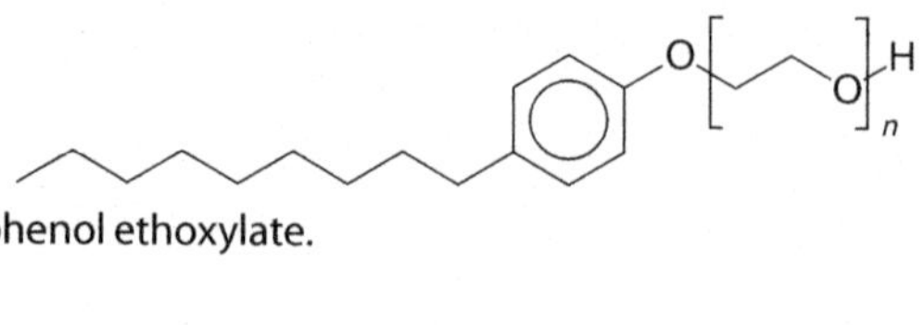

Figure 1 Nonyl phenol ethoxylate.

Table 5 Inhibitive water-based drilling fluid [8]

Compound	Amount
Sodium chloride	24%
CLAY GRABBER™	$0.5\,\mathrm{lbs\,bbl^{-1}}$
CLAY SYNC™	$2.0\,\mathrm{lbs\,bbl^{-1}}$
GEM™	2%
Polyanionic cellulose	$2\,\mathrm{lbs\,bbl^{-1}}$
FILTER CHEK™	$2.0\,\mathrm{lbs\,bbl^{-1}}$
BARAZAN®	$1.0\,\mathrm{lbs\,bbl^{-1}}$
Potassium hydroxide	$0.5\,\mathrm{lbs\,bbl^{-1}}$
BARITE®	

2.8.1 Environmental Aspects

Minimizing the environmental impact of the drilling process is a highly important part of drilling operations to comply with environmental regulations that have become stricter throughout the world. In fact, this is a mandatory requirement for the North Sea sector. The drilling fluids industry has made significant progress in developing new drilling fluids and ancillary additives that fulfill the increasing technical demands for drilling oil wells. These additives have very little or no adverse effects on the environment or on drilling economics.

New drilling fluid technologies have been developed to allow the continuation of oil-based performance with regard to formation damage, lubricity, and wellbore stability aspects and thus penetration rates. These aspects were greatly improved by incorporating polyols or silicates as shale inhibitors in the fluid systems.

Despite their environmental acceptance, conventional water-based drilling muds (WBMs) exhibit major deficiencies relative to OBMs/pseudo-oil-based drilling muds (POBMs) with regard to their relatively poor shale inhibition, lubricity, and thermal stability characteristics. To overcome those deficiencies, specific additives may, however, be added into the WBM compositions to deliver properties close to OBMs/POBMs, performance while minimizing the environmental impact.

Consequently, to meet the new environmental regulations while extending the technical performance of water-based drilling fluids, a new generation of water-based fluids, also called inhibitive drilling fluids was developed to compete against OBMs. Also, to minimize the formation damage, new types of nondamaging drilling fluids, called drill-in fluid, have been developed to drill the pay zone formations.

Polyols-based fluids contain a glycol or glycerol as a shale inhibitor. These polyols are commonly used in conjunction with conventional anionic and cationic fluids to provide additional inhibition of swelling and dispersing of shales. They also provide some lubrication properties.

Sodium silicates or potassium silicates are known to provide levels of shale inhibition comparable to that of oil-based drilling muds (OBMs). This type of fluids is characterized by a high pH (>12) for optimum stability of the mud system. The inhibition properties of such fluids are achieved by the precipitation or gelation of silicates on contact with divalent ions and lower pH in the formulation, providing an effective water barrier that prevents hydration and dispersion of the shales.

2.9 Foamed Types

An aqueous foam may replace a not foamed well fluid [9]. Namely, in some cases, the density of an aqueous fluid that is used is too high for the type of rock, creating an excessive hydrostatic pressure compared to the fracturing resistance of the rock zones through which the drilling passes. The choice is then made to use aqueous fluids with densities that are reduced by the introduction of gas, and it is attempted to make the thus lightened fluid homogeneous by creating the most stable foam possible so that it will have at least adequate power for cleaning the cuttings.

It has been shown that it is possible to control the production of a stable foam from a composition that is optimized in terms of surfactant concentration if a polymer or a copolymer with a charge that is opposite in sign to that of the charge of the surfactant is combined with such a surfactant.

So, a cationic polymer or copolymer is combined with an anionic surfactant or else, an anionic polymer or copolymer is combined with a cationic surfactant. Anionic and cationic surfactants are summarized in Table 6.

Also polymeric materials of these types may be used. For example, the addition of an anionic cellulose natural polymer such as carboxymethyl cellulose (CMC) with a cationic surfactant such as dodecyltrimethylammonium bromide makes it possible to form a foam because of between-charge interactions, with a surfactant concentration of less than $5 \, \text{m mol} \, \text{l}^{-1}$. The foam that is formed with a concentration of $0.5 \, \text{m mol} \, \text{l}^{-1}$ of dodecyltrimethylammonium bromide and 750 ppm of CMC is very stable. It has been verified that an excessive surfactant concentration hampers the formation of a foam, very likely due to the excessively strong interactions between the polymer and the surfactant [9].

Table 6 Anionic and cationic surfactants [9]

Anionic types	Cationic types
Soaps of alkali metals	Alkylamine salts
N-Acylamino acids	Quaternary ammonium salts
N-Acylglutamates	Alkyltrimethyl compounds
N-Acylpolypeptides	Trimethylammonium compounds
Alkylbenzenesulfonates	Alkyldimethyl benzylammonium compounds
Paraffin sulfonates	Pyridinium compounds
Lignosulfonates	Imidazolinium compounds
Sulfosuccinic derivatives	Quinolinium compounds
Alkyl taurides	Piperidinium compounds
Alkyl sulfates	Morpholinium compounds
Alkyl ether sulfates	
Alkyl phosphates	

2.10 Volcanic Ash Containing Types

Bentonite mud is widely used for surface hole drilling, but it has serious technical limitations due to its poor tolerance to monovalent and divalent salts, undesirable mud solids, cement contamination, pH changes, and temperature changes above 100 °C [10]. Also, this mud system has strong interactions with subsurface formations such as clay–rich formations and reactive shale.

Due to the high dissolution capacity of fresh water used in bentonite mud formulations, it often leads to hole enlargement and loss of circulation.

A WBM composition with improved properties relative to tolerance of high salt content, cement, lime, and temperatures has been described. Such a composition includes volcanic ash, water, a viscosifier, a pH buffer, and a bifunctional mud additive that acts both as a viscosity enhancer and also as a fluid loss additive [10].

Volcanic ash predominantly consists of silica, aluminum oxide, calcium oxide, ferric oxide, and magnesium oxide and can be found all over the world. Volcanic ash is substantially different from other types of ash, such as fly ash, which primarily contains silicon dioxide and lime. There can be some variations in the composition of the volcanic ash depending on the locality, area, and country of deposition.

3. SPECIAL ADDITIVES FOR DRILLING FLUIDS

Subsequently, special additives for drilling fluids are discussed. These additives may be used also for other operations than drilling with water-based

Table 7 Water-based drilling muds

Compound	References
Glycol-based	[11]
Alkali silicates	[12, 13]
Poly(acrylamide), CMC	[14]
CMC, zinc oxide	[15]
AAm copolymer, poly(propylene glycol) (WBM)	[16]

drilling fluids, but have been in particular described in the literature in context with drilling fluids.

Commercial products are listed in the literature. These include bactericides, corrosion inhibitors, defoamers, emulsifiers, fluid loss and viscosity control agents, and shale control additives. WBMs are summarized in Table 7.

3.1 Rheology Control

3.1.1 Compositions with Improved Thermal Stability

To avoid the problems associated with viscosity reduction in polymer-based aqueous fluids, formates, such as potassium formate and sodium formate are commonly added to the fluids to enhance their thermal stability. However, this technology of using the formates is very expensive. The thermal stability of polymer-based aqueous fluids can be improved without the need of using formates [17].

The stability of a wellbore treatment fluid may be maintained at temperatures up to 135–160 °C (275–325 °F). Various poly(saccharide)s may be included in the fluid. The apparent viscosities of drilling fluids containing xanthan gum and PAM before and after rolling at 120 °C are shown in Table 8.

3.1.2 Carboxymethyl Cellulose

Sodium CMC and PAC, a CMC which has a degree of substitution usually greater than 1.0, are two of the more widely used anionic polymers which serve to control the viscosity and the filtration rates [3].

A higher degree of substitution CMC, e.g., in PAC, offers good fluid loss reduction in an electrolyte-containing system with a smectite-type clay such as bentonite. However, the effectiveness of CMC, being a polyelectrolyte, as a viscosity builder has its limitations, as its effectiveness decreases with increasing electrolyte concentration. Thus, regular CMC is mostly suitable for electrolyte-poor drilling fluid compositions, such as freshwater-based drilling fluid compositions. Although CMCs and PACs with a high degree

Table 8 Apparent viscosity before and after rolling [17]

Composition	Before η (cP)	After η (cP)
Brine/XC	13	3
Brin/PA	8.5	6
Brine/Filtercheck	4	4
Brine/FLC/XC	16	10.5
Brine/FLC/PA	14.6	9
Brine/XC/CLAYSEAL	12.5	3
XC/PA	30	28.5
XC/PA/FLC	38.5	16.5
XC/PA/FLC/CLAYSEAL	34	28
XC/PA/FLC/CLAYSEAL/Barite	38.5	38.5

of polymerization are used as viscosifiers, the regular grades do not have the good suspension carrying properties (high low-shear viscosity), which are needed to bring the cuttings efficiently to the surface.

Alternatively, xanthan gum is employed as viscosifier and suspending agent [3]. Xanthan gum has very suitable rheological properties. It forms a gel within a short period of time when drilling circulation is slow or interrupted. This enables immobilization of dispersed solids in the fluid composition. After circulation is resumed, the gel easily transforms into a flowing fluid, thereby maintaining a good dispersion of the solids contained in the fluid composition.

However, xanthan gum is relatively expensive. Moreover, it is only stable at temperatures below about 120 °C, which makes it less suitable for drilling at temperatures exceeding 120 °C. Furthermore, many xanthan grades contain very fine insoluble material, usually residues from the fermentation production process. These insoluble materials are undesired for drilling operations, as they cause, e.g., more difficult hole cleaning. Only the more expensive xanthan grades do not have these insoluble parts [3].

In most drilling operations, the drilling fluid composition experiences a high shear. This is particularly advantageous if CMC is used in such fluid compositions, as specific CMC formulations are gelling, when exposed to a high shear. Applying a high shear improves the gelling properties of CMC considerably [3].

The gelling properties of CMC can also be improved by a heat treatment. Preferably, the CMC is treated at 70 °C or higher. A drilling fluid composition has described that can be used up 140 °C. However, above this temperature the CMC decomposes. In comparison to fluid compositions that are using xanthan gum, which deteriorates above 120 °C, deeper drilling operations at higher temperatures can be carried out.

3.1.3 Quaternary Nitrogen Containing Amphoteric Water-Soluble Polymers

A rheological control agent for water-based drilling fluids has been described [18]. This is an amphoteric polymer containing both cationic groups and anionic groups. The cationic groups are quaternary ammonium groups. The amphoteric polymers are based on quaternary nitrogen containing amphoteric carboxymethyl cellulose.

Amphoteric polymers show unusual solution properties. Charge balanced polyampholytes often are more soluble and show higher viscosities in salt than in pure water solution. Therefore, amphoteric polymers have found utility as water and brine viscosifiers and as brine drag reduction agents. In these applications, the unusual interplay of positive and negative charges on the same group or backbone, in between the chains or external electrolytes play an important role. These charge interactions in the different electrolytic environments have a large part in determining the resulting viscosity of the solutions.

Typical quaternary ammonium salts that can be used include quaternary nitrogen containing halides, halohydrins, and epoxides. Also, the quaternary ammonium salt may contain hydrophobic groups. Exemplary ammonium salts are summarized in Table 9. Organic quaternary ammonium salts are shown in Figure 2. Anionic monomers are shown in Figure 3.

The quaternization can be achieved by a two-step synthesis of aminating the polysaccharide by the reaction with an aminating agent, such as an amine halide, halohydrin, or epoxide followed by quaternizing the product by the reaction with quaternizing agents [18].

In order to regulate the charge density and the hydrophilic-hydrophobic balance of the amphoteric synthetic polymer, nonionic monomers can be copolymerized with the cationic and anionic monomers. Some nonionic monomers such as AAm and maleic acid anhydride may yield anionic groups by becoming hydrolyzed during or after the polymerization [18].

3.1.4 Silicone Resins

A silicone resin–based composition may act as a fluid loss control additive for water-based drilling fluids and is nondamaging [19]. The composition is stable in saturated salt at 120 °C and also at elevated pressure. Solid particles of a silicone resin with a glass transition temperature of more than 70 °C are

Table 9 Quaternary ammonium compounds and other monomers [18]
Salts

3-Chloro-2-hydroxypropyl dimethyloctadecyl ammonium chloride
3-Chloro-2-hydroxypropyl-dimethyloctyl ammonium chloride
3-Chloro-2-hydroxypropyl trimethyl ammonium chloride
2-Chloroethyl trimethyl ammonium chloride
2,3-Epoxypropyl trimethyl ammonium chloride
3-Chloro-2-hydroxypropyl dimethyidodecyl ammonium chloride
3-Chloro-2-hydroxypropyl dimethyltetradecyl ammonium chloride
3-Chloro-2-hydroxypropyl dimethylhexadecyl ammonium chloride
Diallyl dimethyl ammonium chloride

Cationic monomers

Acryloyloxyethyl-trimethyl ammonium chloride
Methacrylamido propyl trimethyl ammonium chloride
3-Acrylamido-3-methyl-butyl-trimethyl ammonium chloride

Anionic monomers

Acrylic acid
Methacrylic acid
2-Acrylamido-2-methylpropane sulfonic acid
Vinylsulfonic acid
Styrene sulfonic acid

Nonionic monomers

AAm
Maleic acid anhydride
Methyl acrylate
Ethyl acrylate
Hydroxyethyl acrylate
Hydroxypropyl acrylate
Butyl acrylate
Vinylacetate
Styrene

used. Wellbore fluids are prepared by first forming the solid silicone resin particulate with the desired particle size distribution from a solution of the silicone resin, and then dispersing the silicone resin into a liquid carrier.

3.1.5 Friction Reducers

During certain operations in subterranean wells, aqueous treatment fluids are often pumped through tubular goods [20]. A considerable amount of energy may be lost due to friction between the aqueous treatment fluid

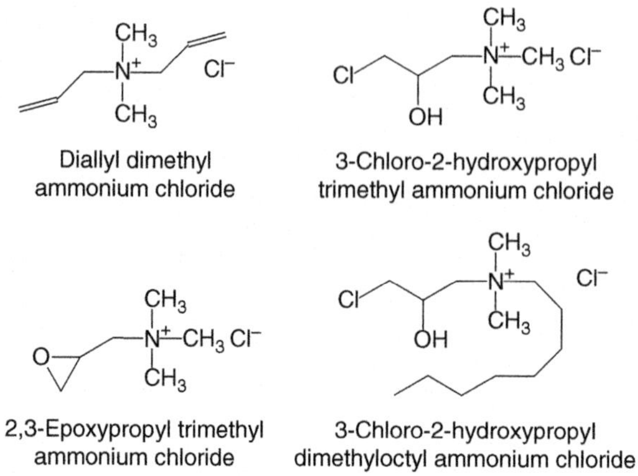

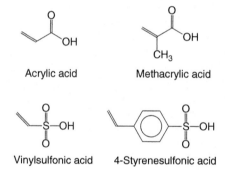

Figure 2 Organic quaternary ammonium salts.

Acrylic acid   Methacrylic acid

Vinylsulfonic acid   4-Styrenesulfonic acid

Figure 3 Organic acids.

in turbulent flow. In order to reduce these energy losses, friction reducing polymers can be included in aqueous treatment fluids.

Partially hydrolyzed polyacrylamide or a copolymer from AAm and acrylic acid (AA) or other monomers can be used as friction reducers [20]. Other comonomers to AAm are AA, 2-acrylamido-2-methylpropane sulfonic acid, N,N-dimethylacrylamide, vinylsulfonic acid, N-vinyl acetamide, N-vinylformamide, itaconic acid, methacrylic acid, and diallyl dimethyl ammonium chloride.

A nontoxic environmentally acceptable friction pressure reducer consists of a mixture of a copolymer of AAm and dimethylaminoethyl acrylate quaternized with benzyl chloride and a stabilizing and dispersing homopolymer of ethanaminium,N,N,N-trimethyl-2-[(1,oxo-2-propenyl)oxy]-chloride [21]. The friction pressure reducer can be prepared as follows [21]:

Preparation II–1: A three-necked flask equipped with a stirrer, anl inlet nitrogen purging tube and a condenser was charged with 52 g of deionized water. To the water, 18 g of ammonium sulfate, 7.5 g of sodium sulfate, and 2 g of glycerol were added. The salt solution formed was continuously agitated. 15.35 g of AAm, 1.15 g of dimethylaminoethyl acrylate quaternized with benzyl chloride, and 4 g of a stabilizing and dispersing homopolymer of ethanaminium,*N,N,N*-trimethyl-2-[(1,oxo-2-propenyl)oxy]-chloride were then added to the salt solution while the solution was being stirred. 0.1 g of naphthol ethyl ether surfactant and less than 0.1 g of azo initiator, i.e., 1H-imidazole,2,2′-(azobis-methylethylidene)bis(4,5-dihydro-dihydrochloride) were then added to the stirring solution. The solution was maintained at a temperature of from about 40 to 45 °C with continuous nitrogen purging. When the oxygen was completely displaced, the polymerization initiated as indicated by the increasing viscosity of the solution and the production of exothermic heat. The solution was held at in the range of 75-80 °C for about 2 h. Thereafter, the reaction was terminated by raising the temperature to 95 °C and holding that temperature for 1 h. ∎

It has been demonstrated that the friction pressure reducer provides an excellent friction pressure reduction in fresh water, a 2% solution of potassium chloride, a 10% solution of sodium chloride, and a synthetic brine solution [21].

3.1.6 Deflocculants

Controlling the viscosity of WBMs or mud systems has traditionally been done with lignosulfonate deflocculants and thinners [1].

Such low molecular weight, heavily sulfonated polymers are believed to aid in coating clay edges in the subterranean formation with a lasting or effectively permanent negative charge. Some alkaline material, such as, e.g., caustic soda or potash, is typically added to achieve a pH of 9.5-10. This pH environment is believed to aid the solubility and activation of the portion(s) of the lignosulfonate molecules that interact with the clay. These portions are believed to be the carboxylate and phenolate groups on the lignosulfonate.

Lignosulfonates are obtained from byproducts of the spent acid process used to separate cellulose from wood in the pulp industry. The pulp industry has begun to turn away from the spent acid process in recent years in favor of another process that does not have a lignosulfonate byproduct.

Consequently, the drilling fluid industry has begun efforts to find a substitute for lignosulfonates in drilling fluids [1].

Also, increasingly, there is an interest in and need for deflocculants and thinners that can work effectively at lower pH of 8-8.5, in freshwater and salt WBMs, and at higher temperatures ranging upwards to about 230 °C, while also being environmentally compatible.

Various methods for the modification of lignosulfonates have been described, e.g., by condensation with formaldehyde [22] or modification with iron salts [23]. It has been found that chromium-modified lignosulfonates, as well as mixed metal lignosulfonates of chromium and iron, are highly effective as dispersants and therefore are useful in controlling the viscosity of drilling fluids and in reducing the yield point and gel strength of the drilling fluids. Because chromium is potentially toxic, its release into the natural environment and the thereof is continuously being reviewed by various government agencies around the world.

Therefore, less toxic substitutes are desirable. Less toxic lignosulfonates are prepared by combining tin or cerium sulfate and an aqueous solution of calcium lignosulfonate, thereby producing a solution of tin or cerium sulfonate and a calcium sulfate precipitate [24].

3.1.7 Ferrous Sulfate

The viscosity of a fluid normally decreases with increasing temperature, however certain polymer additives or deflocculating agents may reduce or even reverse, this behavior.

Water-based drilling fluids containing poly(N-vinyl-2-pyrrolidone/ sodium 2-acrylamido-2-methylpropane sulfonate) provide an excellent fluid loss control in hostile environments.

However, this polymer enhances the viscosity in drilling muds so much that enough of this material cannot be incorporated into high solids muds to attain a desired degree of fluid loss control. It has been found that the viscosity of poly(N-vinyl-2-pyrrolidone/sodium 2-acrylamido-2-methylpropane sulfonate) can be reduced by using a ferrous sulfate as additive [25].

3.1.8 Biopolymers

Biopolymers have been used in nondamaging drilling fluids [26]. Biopolymers can be selected from nonionic and ionic poly(saccharide)s.

Biopolymers of the polysaccharide type are hydrosoluble polymers and are produced by bacteria or fungi. Their function is to increase the viscosity of the fluid in order to keep filtrate reducers and drilled solids suspended both during the drilling period and during possible stop in the fluid circulation.

Actigum® CS 6 is a branched homopolysaccharide produced by a fungus of the Sclerotium type by means of an aerobic fermentation process. Here, the main chain of scleroglucan consists of residues of the β-D-glucopyranosyl type with bonds (1-3), whereas the branchings consist of D-glucopyranosyl residues with β-bonds (1-6) every three glucoside groups.

Scleroglucan is produced by the fermentation of a fungus of the *Sclerotium* type. Scleroglucan, like xanthan, is a hydrophilic colloid which has the property of thickening and stabilizing water-based systems by conferring on them a particularly high viscosity. The viscosity of a scleroglucan solution is virtually independent of the pH between 1 and 12.5 and of the temperature up to a limit of 130 °C. The viscosity is also highly independent of the salinity [27].

It has been found that by using nonrefined scleroglucan as the viscosity agent, it was possible to reduce the concentration of the other constituents of the mud or even to suppress them [27]. In fact, the introduction of scleroglucan as main constituent permits the preparation of a WBM containing a minimum of constituents ensuring its functions in a satisfactory manner in a large number of cases, especially if the scleroglucan used contains all or part of the mycelium which produced it. The presence of this mycelium in the mud improves some of its functions. Such a mud appears to preserve the subterranean formations better than a mud containing a refined scleroglucan. The subterranean formations are then less sensitive to the effect of erosion and mechanical destruction.

Welan gum produced by the *Alcaligenes* species. This is a heteropolysaccharide similar to scleroglucan but with residues of the α-L-rhamnopyranosyl or α-L-mannopyranosyl type.

Xanthan gum is an anionic branched heteropolysaccharide produced by the bacterium *Xanthomonas campestris*, characterized by a main chain consisting of D-glucopyranoside residues with lateral chains consisting of pyruvate and acetate groups [26]. Xanthan has a low sensitivity to significant variations in temperature, pH, or salinity [27].

A formulation composition is shown in Table 10.

3.1.9 Biodegradable Thinners

WBMs and compositions that are biodegradable contain a polyamide-based copolymer having at least one grafted side chain from AA and

Table 10 Formulation composition for drilling fluids [26]

Compound	No. 1	No. 2
Water	1000 ml	900 ml
Xanthan gum	16 g	–
Starch	16 g	–
Crosslinked cellulose fibers	–	118 g
CaCO$_3$	–	90 g

Table 11 Polyamides and vinyl monomers suitable for grafting [28, 29]

Polyamide type	Vinyl compound
Casein	Vinyl ether
Gelatine	Acrylic acid
Collagen	Methacrylic acid
Bone glue	2-Ethylacrylic acid
Blood albumin	2-Propylacrylic acid
Soy protein	Vinylacetic acid
	Vinylphosphonic acid
	Crotonic acid
	Isocrotonic acid
	Maleic acid
	Fumaric acid
	Itaconic acid
	Citraconic acid

acrylo amido propane sulfonate (AMPS) [1]. It is further believed that the AMPS functional groups impart a higher temperature stability and greater functionality at a wider pH range.

Typical representatives of the polyamide component are natural polyamides, like caseins, galantines, collagens, and soy proteins [28, 29]. The monomers which are suitable for grafting are given in Table 11. Some vinyl monomers are shown in Figure 4.

Such a thinner/deflocculant can be used both in a freshwater drilling fluid base, as well as in a brine drilling fluid base. Further, it is effective even at a near neutral pH of 8.0-8.5, while still being effective at a higher pH [1].

The drilling fluids maintain a satisfactory rheology for drilling at temperatures as high as 200-230 °C, but are also useful at lower temperatures, such as temperatures as low as 4 °C.

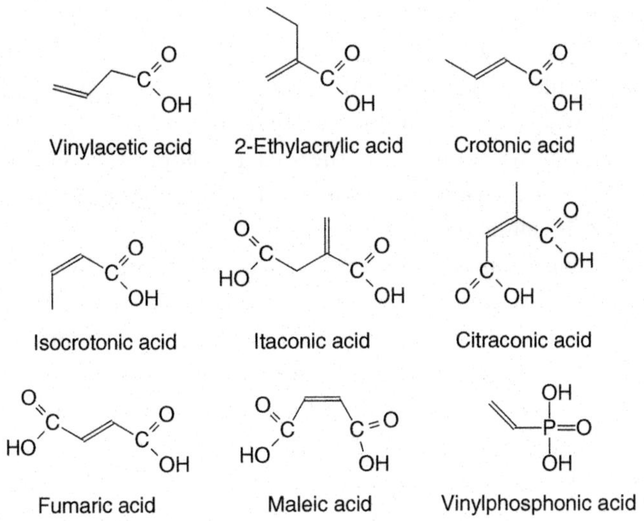

Figure 4 Vinyl monomers.

Further, the system does not contain chromium, which is commonly used with lignosulfonate thinners. Therefore such a composition is more environmentally friendly [1].

3.1.10 High-Temperature Formulation

Because of their better thermal stability as compared to water-based fluids, oil-based fluids typically have been used in high-temperature applications [30]. However, as the environmental impact of the disposal of these spent slurries, and the drilled cuttings carried in these slurries, has become increasingly scrutinized, water-based fluids have become more and more the fluid of choice in the industry.

Water-based fluids are also preferable in high-pressure applications, such as deep wells, because oil-based fluids are more compressible than water-based fluids. The increased compressibility results in an increased viscosity. Another advantage for water-based drilling fluids is the safety aspect in well-control situations. Gases are much more soluble in oil; therefore, an unanticipated influx of gas into the well cannot be detected as well at the surface in oil-based drilling fluids until it is near the surface and very dangerous.

Since gas is only sparingly soluble in water-based drilling fluids, such an influx can be detected more easily, the well shut-in, and the influx can be more safely handled.

The major difficulty in a typical WBM at higher temperatures is that it may degrade and will become too viscous to be circulated easily. This circulation difficulty arises because the clays used in the muds are susceptible to a temperature-induced gelation at temperatures of already 120 °C.

A formulation with improved thermal rheological stability has been presented [30]. The major components are clay, an inorganic salt, and parenchymal cell cellulose.

The rheological stability of a mud can be monitored by measuring its yield point and gel strength, according to standard drilling fluid tests [31].

Water-based drilling fluids which display a rheological stability throughout a wide temperature range have been described. This rheological stability enables the fluid to carry drilled cuttings efficiently at ambient temperatures. It also provides a sufficiently fluid viscosity at higher bottom hole temperatures to provide an ease of circulation downhole. The drilling fluid comprises three components: clay, inorganic salt such as a chloride salt or a sulfate salt, and parenchymal cell cellulose [32].

Parenchyma cells constitute the bulk of the soft parts of plants. It has been discovered that the cellulose from cell walls of parenchymal cells of certain plants, especially of sugar beets and citrus, possesses a unique morphology together with unique chemical and physical properties. Suspensions of such parenchymal cell cellulose, when isolated from such plants, exhibit a highly beneficial rheological behavior [33]. Further, fruit or vegetable cells with an exclusively parenchymal cell wall structure can be treated with a mild process to form highly absorbent microfibers [34].

3.1.11 Low-Temperature Formulation

Antifreeze agents are occasionally added to reduce the freezing point of the drilling fluid itself [35, 36]. Such a water-based drilling fluid is composed of water, clay or polymer, and a polymeric glycerol.

More recently, for low-temperature drilling, antifreeze agents such as poly(vinylpyrrolidone), quaternary ammonium salts, or antifreeze fish proteins have been proposed [37].

3.2 Fluid Loss Control

3.2.1 Membrane Formation

An osmotic membrane is a porous material with pores small enough to prevent solute flow but large enough to allow the solvent to flow [38]. A chemical imbalance is created by placing fluids with different concentrations (molalities) of solute on opposite sides of the membrane. The flow of solvent through the membrane into the solute-rich solution causes the solution volume to expand, which creates a pressure. The volume of solvent will increase until the chemical potential of the solute-rich solution in the membrane equals the potential of the solution containing pure solvent. This pressure created by increases in solvent volume is called the osmotic pressure.

In order to increase the wellbore stability, it is possible to provide formulations for water-based drilling fluids, which can form a semi-permeable osmotic membrane over a specific shale formation [39]. This membrane allows a comparatively free movement of water through the shale, but it significantly restricts the movement of ions across the membrane and thus into the shale.

The method of membrane formation involves the application of two reactants to form in situ a relatively insoluble Schiff base, which deposits at the shale as a polymer film. This Schiff base coats the clay surfaces to build a polymer membrane.

The first reactant is a soluble monomer, oligomer, or polymer with ketone or aldehyde or aldol functionalities or precursors to those. Examples are carbon hydrates, such as dextrin, linear of branched starch. The second reactant is a primary amine. These compounds are reacting by a conden-sation reaction to form an insoluble crosslinked polymerized product. The formation of a Schiff base is shown in Figure 5.

Figure 5 illustrates the reaction of a dextrine with a diamine, but other primary amines and poly(amine)s will of course react in the same way. Long-chain amines, diamines, or poly(amine)s with a relatively low amine ratio may require supplemental pH adjustment using materials such as sodium hydroxide, potassium hydroxide, sodium carbonate, potassium carbonate, or calcium hydroxide [39]. The Schiff base formed in this way must be essentially insoluble in the carrier brine in order to deposit a sealing membrane on the shale during the drilling of a well.

The polymerization and precipitation of the osmotic membrane on the face of the exposed rock significantly retards water or ions from moving into or out of the rock formation, typically shale or clay. The ability to form an osmotic barrier results in an increased stability in the clays or minerals, which combine to make the rock through which the borehole is being drilled [39].

Water-based drilling fluids have been described that can form a semi-permeable osmotic membrane over a shale formation to increase the wellbore stability [40]. The membrane allows a relatively free movement of water out of the shale, but significantly restricts the movement of ions across the membrane and into the shale or clay.

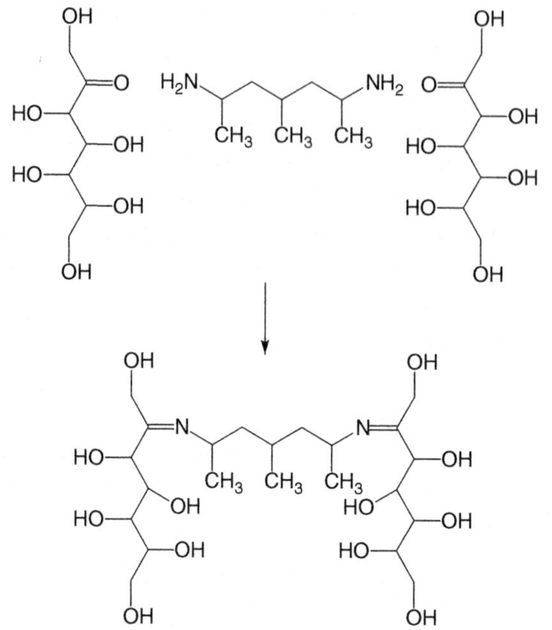

Figure 5 Formation of a Schiff base [39].

In particular, the stability of the shale formation with a water-based drilling fluid can be increased by [40]:

1. delivering to the shale formation a drilling fluid formulated to include an aqueous fluid;
2. a first reactant which is a soluble monomer, oligomer, or polymer with exposed ketone, aldehyde, or aldol groups or with groups which can be shifted to ketone or aldehyde functionality;
3. A second reactant which is a primary amine, diamine, or polyamine which by condensation reaction forms a semi-soluble or precipitated filming product with the first reactant.

Variables that are of significance in producing a membrane include the primary polymer content and molecular weight, the crosslinking amine content, and the pH of the solution.

By carefully selecting the primary polymer and the crosslinking amine, the crosslinking and polymerization and precipitation of components effectively occur. Reactants are summarized in Table 12.

The Schiff base formed by the reaction of the reactants must be partially soluble or insoluble in the carrier brine in order to form a sealing membrane on shale or other formation exposed during drilling of a well.

Table 12 Reactants [40]

Carbonyl compounds	Amine compounds
Starch	Hexamethylene diamine
Dextrin	Ethoxylated alkyl ether amine
Methyl glucosides	Propoxylated alkyl ether amine
Corn syrup	Polyoxy propylene diamine
Malto-dextrin	
Molasses	
Sugar	
Cellulose	

In the course of reaction, a very significant darkening is associated with the classic Maillard reaction and provides an evidence of a Schiff base polymerization [40]. Actually, the Maillard reaction causes the browning of food in the course of frying. It results from a chemical reaction between an amino acid and a reducing sugar [41].

In argillaceous formations using water-based drilling fluids, the stability can be improved by using long-chain fatty acids that can self-assemble in their ionized state, or methyl silanetriol and related compounds [42].

The condensation polymerization of silicic acid involves the reaction of a silicate ion with an unionized silanol group and ends in an polymerization-depolymerization equilibrium [42].

A range of phenols were evaluated for their membrane generation capacity at about 25 °C and using test solutions at about 15 MPa. The compounds include 2-naphthol which precipitates as a solid. Tests were conducted with 2-naphthol of 10% concentration and pH of 11.8-12. The water activity of the solutions can be reduced with 12% sodium chloride. Membrane efficiencies of 65% were obtained [42].

Actually, the membrane efficiency of shales when interacting with different WBMs and OBMs was assessed [43]. Pressure transmission tests were used to measure the membrane efficiency using cations and anions at different concentrations, i.e., water activities.

The membrane efficiencies of shales, when exposed to salt solutions, are low-ranging from 0.18% to 4.23%. Correlations between the membrane efficiency and other shale properties have been found. For example, the membrane efficiency of shales is directly proportional to the ratio of the cation exchange capacity and permeability of shales. Thus, higher cation exchange capacities and lower permeabilities correlate very well with higher membrane efficiencies.

In addition, the ratio of the size of the hydrated ions to the size of the shale pore throat determines the ability of a shale to restrict solutes from entering the pore space and thus controls its membrane efficiency. So, the formulation of drilling fluids can be tailored in terms of the types of cation and anion used in a water based fluid [43].

Glycosides

The advantage of using glycosides in the internal phase is that much of the concern for the ionic character of the internal phase is no longer required. If water is limited in the system, the hydration of the shales is greatly reduced.

The reduced water activity of the internal phase of the mud and the improved efficiency of the shale is an osmotic barrier if the glycoside interacts directly with the shale. This helps lower the water content of the shale, thus increasing rock strength, lowering effective mean stress, and stabilizing the wellbore [44].

Methyl glucosides also could find applications in water-based drilling fluids and have the potential to replace OBMs [45]. The use of such a drilling fluid could reduce the disposal of oil-contaminated drilling cuttings, minimize health and safety concerns, and minimize environmental effects. Materials for inverted emulsion drilling fluids are shown in Table 13.

3.2.2 Polydrill

Polydrill is a sulfonated polymer for filtration control in water-based drilling fluids [57]. Tests demonstrated the product's thermal stability up to 200 °C and its outstanding electrolyte tolerance. Polydrill can be used in NaCl-saturated drilling fluids as well as in muds containing 75,000 ppm of calcium or 100,000 ppm of magnesium. A combination of starch with Polydrill was

Table 13 Other materials for inverted emulsion drilling fluids

Base material	References
Ethers of monofunctional alcohols	[46]
Branched didecyl ethers	[47, 48]
α–Sulfofatty acids	[49]
Oleophilic alcohols	[50–52]
Oleophilic amides	[53]
Hydrophobic side chain poly(amide)s (PAs) and sodium poly(acrylate) or poly(acrylic acid)	[54]
Poly(ether amine)	[55]
Phosphate ester of a hydroxy polymer	[56]

used successfully in drilling several wells. The deepest hole was drilled with 11-22 kg m^{-3} of pregelatinized starch and 2.5-5.5 kg m^{-3} of Polydrill to a depth of 4800 m. Field experience with the calcium tolerant starch/Polydrill system useful up to 145 °C has been discussed in detail [58].

In dispersed muds (e.g., lignite or lignosulfonate), minor Polydrill additions result in a significantly improved high-temperature, high-pressure filtrate. Major benefits come from a synergism of polymer with starch and poly(saccharide)s. The polymer exerts a thermal stabilizing effect on those polymers.

In conventional or clay-free drilling and completion fluids, Polydrill can be used by itself or in combination with other filtrate reducers for various purposes [59]. Handling and discharge of the product as well as the waste mud created no problem in the field.

3.2.3 Polymer of Monoallylamine

A water-soluble polymer of monoallylamine, c.f., Figure 6, can be used in conjunction with a sulfonated polymer, such as a water-soluble ligno-sulfonate, condensed naphthalene sulfonate, or sulfonated vinyl aromatic polymer, to minimize the fluid loss from the slurry during well cementing operations [60, 61]. The polymer of monoallylamine may be a homopolymer or a copolymer and may be crosslinked or uncrosslinked.

These components react with each other in the presence of water to produce a gelatinous material that tends to plug porous zones and to minimize premature water loss from the cement slurry to the formation.

In addition, the filtration value becomes significantly reduced after high-temperature aging when a crosslinked copolymer is used in a standard water-based drilling fluid formulation [62].

3.2.4 Combination of Nonionic and Ionic Monomers

Many additives for water-based fluids have been found to effectively provide fluid loss control, increase viscosity, and inhibit drill solids.

The fluids usually contain little or no calcium chloride. However, when the calcium chloride concentration increases in the fluids, the effectiveness

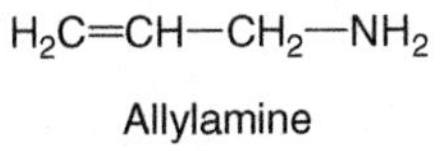

$$H_2C=CH-CH_2-NH_2$$

Allylamine

Figure 6 Monoallylamine.

of these additives, especially for maintaining rheology and water loss control, decreases significantly [63].

Terpolymers and tetrapolymers have been proposed as fluid loss additives for drilling fluids [64, 65]. The constituent monomers are a combination of nonionic monomers and ionic monomers.

Tetrapolymers using N-vinyl-2-pyrrolidone, acrylamido propane sulfonic acid, AAm, and AA as monomers can be prepared by radical polymerization in solution, suspension, or emulsion [66]. However, the method of bulk solution polymerization is preferred [67]. As radical initiator $2,2'$-azobis-(N,N'-dimethyleneisobutyramidine) dihydrochloride is used. In addition, a chelating agent, EDTA, is added in a stoichiometric excess. The reaction is carried out with an inert gas at up to 60 °C.

Also, the nonionic monomer can be AAm, N,N-dimethylacrylamide, N-vinyl-2-pyrrolidone, N-vinyl acetamide, or dimethyl amino ethyl methacrylate. Ionic monomers are AMPS, sodium vinylsulfonate, and vinylbenzene sulfonate. The terpolymer should have a molecular weight between 0.2 and 1 MD.

A formulation consisting of AMPS, AAm, and itaconic acid has been proposed [68]. Such polymers are used as fluid loss control additives for aqueous drilling fluids and are advantageous when used with lime muds or gypsum-based drilling muds containing soluble calcium ions.

For seawater muds, another example is a copolymer of 10% AMPS and 90% AA in its sodium salt form [69]. The polymers have an average molecular weight of 50–1000 kD.

A terpolymer from a family of intramolecular polymeric complexes (i.e., polyampholytes), which are terpolymers of AAm-methyl styrene sulfonate-methacrylamido propyl trimethyl ammonium chloride has been reported [70, 71].

A terpolymer formed from ionic monomers AMPS, sodium vinylsulfonate or vinylbenzene sulfonate itaconic acid, and a nonionic monomer, e.g., AAm, N,N-dimethylacrylamide, N-vinyl-2-pyrrolidone, N-vinyl acetamide, and dimethyl amino ethyl methacrylate, is used as a fluid loss agent in oil well cements [72].

The terpolymer should have a molecular weight of 0.2–1 MD. The terpolymer comprises AMPS, AAm, and itaconic acid. Such copolymers also serve in drilling fluids [73].

A tetrapolymer consisting of 40–80 mol.% of AMPS, 10–30 mol.% of vinylpyrrolidone, 0–30 mol.% of AAm, and 0–15 mol.% of acrylonitrile was also suggested as a fluid loss additive [74]. Even at high salt concentrations, these polymers yield high-temperature-stable protective colloids that provide for minimal fluid loss under pressure.

For water-based drilling fluids, water-soluble polymers are adequate, typically PAM. A disadvantage of these polymers is that they have a limited temperature stability.

Polymers based on AAm have been developed, which exhibit effective rheological properties and high temperature/high pressure (HTHP) filtration control at temperatures of 260 °C or more [75].

These are terpolymers composed from AAm, AMPS, or alkali metal salts, and a third monomer which can be an acrylate, N-vinyl lactam, or N-vinylpyridine (NVP) [63, 75]. Examples of copolymers are listed in Table 14.

In order to assist effective seal forming for filtration control, a plugging agent is added [75]. Examples of suitable plugging agents include sized sulfonated asphalt, limestone, marble, mica, graphite, cellulosics, lignins, and cellophanes.

Other additives may be used in the drilling fluid system. Such additives includ shale stabilizers, further filtration control additives, suspending agents, dispersants, anti-balling additives, lubricants, weighting agents, seepage control additives, lost circulation additives, drilling enhancers, penetration rate enhancers, corrosion inhibitors, buffers, gelling agents, crosslinking agents, salts, biocides, and bridging agents [75].

Monomers suitable for crosslinking are N,N'-methylenebisacrylamide, divinyl benzene, allyl methacrylate, or tetraallyl oxethane [76]. These monomers are shown in Figure 7. Crosslinked polymers are effective as high-temperature fluid loss control agents.

Table 14 Examples of copolymers [75]

Monomer 1	mol.%	Monomer 2	mol.%	Monomer 3	mol.%
AMPS	10	AAm	90	–	–
AMPS	20	AAm	80	–	–
AMPS	40	AAm	60	–	–
AMPS	37.5	AAm	50	Acrylate	12.5
AMPS	55	AAm	15	NVP	30
Monomer 1	% w/w	Monomer 2	% w/w	Monomer 3	% w/w
NaAMPS	90	DMAAm	10	–	–

[a]AMPS, 2-acrylamido-2-methyl-1-propane sulfonic acid; AAm, acrylamide; NVP, N-vinyl-2-pyrrolidone; NaAMPS, sodium AMPS; DMAAm, N,N-dimethylacrylamide.

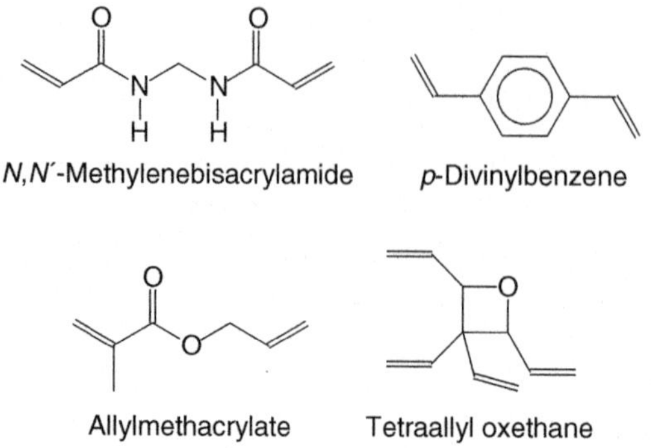

N,N'-Methylenebisacrylamide   p-Divinylbenzene

Allylmethacrylate   Tetraallyl oxethane

Figure 7 Multifunctional monomers.

After static aging up to 260 °C for 16 h, the HTHP filtrate was measured extensively, with results of less than 25 cm^3 min^{-1} [75].

In a CaCl$_2$ water mud, samples containing copolymers of AAm and N-Vinyl-2-pyrrolidone provide a much lower fluid loss than the same with the AAm and sodium-2-acrylamido-2-methanesulfonate copolymer [63].

Similar copolymers with N-vinyl-N-methylacetamide as a comonomer have been proposed for hydraulic cement compositions [77].

The polymers are effective at well bottom hole temperatures ranging from 93 to 260 °C (200-500 °F) and are not adversely affected by brine. Terpolymers of 30-90 mol.% AMPS, 5-60 mol.% of styrene, and residual AA are also suitable for well cementing operations [78]. Copolymer blends for fluid loss are shown in Table 15.

3.2.5 Synthetic Polymeric Fibers

Fluid loss is common in drilling operations. Therefore, drilling fluids are designed to seal porous formations intentionally during drilling [82].

There, a mud cake is created to seal the porous formation. However, in some cases, the loss of the whole slurry into the formation can reach an extent such that no mud cake is created to secure the surface and to create an effective barrier.

In extreme situations, the borehole penetrates a fracture in the formation and most of the drilling fluid is lost. In such a case, the lost circulation zone

Table 15 Copolymer blends for fluid loss [79–81]

Copolymer	Molecular weight (kD)
Acrylamide/vinyl imidazole	100-3000
Vinylpyrrolidone/sodium vinylsulfonate	100-3000

must be sealed in a separate operation. Or, in the worst case, a loss of the well occurs.

A water-based composition can be used for reducing lost circulation while drilling a well is done with an OBM. Such a composition includes an aqueous base fluid, a mixture of coarse, medium, and fine particles, and a blend of long and short fibers [82]. Optionally, the long fibers are rigid and the short fibers are flexible.

For example, the long fibers may consist of a water–insoluble poly(vinyl alcohol) and the short fibers are a water–soluble poly(vinyl alcohol) or an inorganic fiber [82].

The rigid fibers initially create an effective three-dimensional heterogeneous mesh across a fracture. Eventually, this mesh is blocked by the short flexible fibers and the particles. With such a water-based treatment, wider fractures can be treated in comparison to the use in an oil-based treatment.

Synthetic polymeric fibers are used in a drilling mud for controlling the fluid loss in a borehole [83]. A thermal stability of up to about 300°C is desirable. In particular, the fibers are fluoropolymer fibers, in particular made from poly(tetrafluorethylene).

Fluoropolymer fibers are resistant against an atmosphere including CO_2, H_2S, and other chemicals which may present in a borehole or in a surrounding formation.

When using poly(tetrafluorethylene) fibers, the density of about $2.1 \, \mathrm{g \, cm^{-3}}$ of these fibers is different from an average density of for instance $2.6 \, \mathrm{g \, cm^{-3}}$ of drill cuttings. By centrifugation or other methods, a separation of the two components is possible. Also the different viscosity properties of the low friction synthetic polymer fibers on the one hand and other material on the other hand can be used for such a separation. Also different sizes may be used for the separation, for instance by using a strainer [83].

3.2.6 Carboxymethylated Raw Cotton Linters

The addition of carboxymethylated raw cotton linters (RCLs) reduce the fluid loss of water-based drilling fluids [84]. RCL is a naturally occurring ligno–cellulosic material. It is obtained by delinting the leftover fibers from the cottonseed surface after removal of staple fibers. In addition to cellulose, which is a high molecular weight polymer of glucose, raw cotton linters contain lignin and hemicellulose that are also high molecular weight species.

One of the unique attributes of RCLs is that the molecular weights of its polymeric species (cellulose, lignin, and hemicellulose) and the fiber structures of RCLs remain intact during the delinting process. In contrast, isolation of purified celluloses from RCLs and wood pulps involving thermal, mechanical, and chemical means occasions removal of lignin and hemicellulose and substantial molecular weight loss of the recovered cellulose.

The countercations of the carboxymethylated RCLs could be lithium, sodium, potassium, calcium, aluminum, barium, magnesium, ammonium, or a mixture from these compounds.

The carboxymethylated RCLs can be crosslinked with to further improve the performances of the drilling fluids. Examples of ionic crosslinking agents are polyvalent metal ions, such as Al^{3+}, Zn^{2+}, Ca^{2+}, Mg^{2+}, Ti^{4+}, and Zr^{4+}. Covalent crosslinking agents include dichloroacetic acid, trichloroacetic acid, and diahalogenoalkanes.

3.2.7 Poly-anionic Cellulose
A composition containing PAC and a synthetic polymer of sulfonate has been tested for reducing the fluid loss and for the thermal stabilization of a water-based drilling fluid for extended periods at deep well drilling temperatures [85].

3.2.8 Sulfonate
When a sulfonate-containing polymer is added to a drilling fluid containing PAC, the combination reduces the fluid loss. Improved fluid loss is obtained when PAC and the sulfonate-containing polymer, which has a molecular weight of 300-10,000 kD, are combined in a WBM after prolonged aging at 300 °F (150 °C).

3.2.9 Carboxymethyl Cellulose
Certain admixtures of carboxymethyl hydroxyethyl cellulose (HEC) or copolymers and copolymer salts of *N,N*-dimethylacrylamide and 2-acrylamido-2-methyl-1-propane sulfonic acid (AMPS), together with a copolymer of AA, may provide fluid loss control to cement compositions under elevated temperature conditions [86].

3.2.10 Hydroxyethyl Cellulose
HEC with a degree of substitution of 1.1-1.6 has been described for fluid loss control in water-based drilling fluids [87]. An apparent viscosity in

water of at least 15 cP should be adjusted to achieve an API fluid loss of less than 50 ml/30 min. Crosslinked HEC is suitable for high-permeability formations [88, 89].

A derivatized HEC polymer gel exhibited excellent fluid loss control over a wide range of conditions in most common completion fluids. This particular grated gel was compatible with the formation material and caused little or no damage to original permeability [90].

Detailed measurements of fluid loss, injection, and regained permeability were taken to determine the polymer particulate's effectiveness in controlling fluid loss and to assess its ease of removal. HEC can be etherified or esterified with long-chain alcohols or esters. An ether bond is more stable in aqueous solution than is an ester bond [91].

3.2.11 Starch

Starch, c.f., Figure 8, has been traditionally used as raw material for polymeric additives in order to control the fluid loss properties of a drilling mud. The characteristics of the fluid loss of several newly developed starch types have been assessed. Starch products with different contents of amylose were under investigation. Details are shown in Table 16.

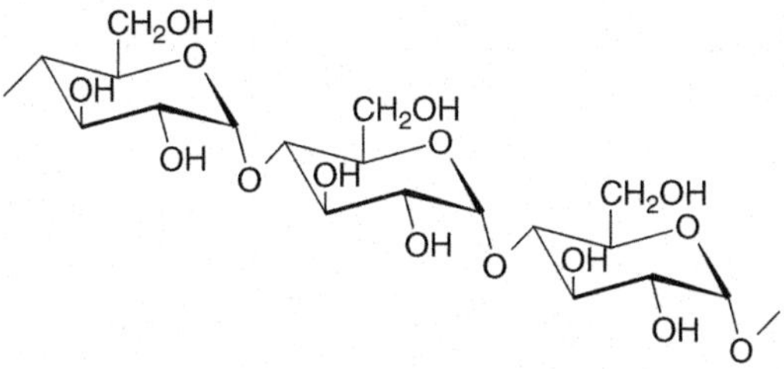

Starch (Amylose)

Figure 8 Starch.

Table 16 Starch products [92]

Starch type	Moisture (%)	Amylose (%)	Molecular weight (kD)
Waxy	12.9	0	20,787
Low amylose	12.7	26	13,000
Intermediate amylose	12.3	50	5115
High amylose	12.2	80	673
Crosslinked high amylose	–	80	–

The products are manufactured by a gelatinization process during reactive extrusion. The extrusion was carried out at 80 bar and 140 °C with a residence time of 3 min. For the crosslinked high amylose type, a chemical was introduced in the course of the extrusion process.

The starches have negligible impurities. No solvent is needed during gelatinization, and further, no waste water is produced as a byproduct. Thus, these types are suitable for environmental sensitive areas.

The presence of most of the starches in a bentonite mud reduce the API filtration of the bentonite mud at room temperature. However, the presence of the chemically modified crosslinked high amylose did not show any improvement in the filtration behavior of a bentonite mud.

The static fluid loss properties have been measured after a thermal treatment at different temperatures. These indicate that the new starch products can be used as fluid loss additives for drilling boreholes having a bottom hole temperature up to 150 °C [92].

The results indicate that some of the starches have static and dynamic fluid loss characteristics similar to or better than those of a widely used modified starch.

3.2.12 Crosslinked Starch

A crosslinked starch was described as a fluid loss additive for drilling fluids [93, 94]. The additive resists degradation and functions satisfactorily after exposure to temperatures of 250 °F (120 °C) for periods of up to 32 h. To obtain crosslinked starch, a crosslinking agent is reacted with granular starch in an aqueous slurry. The crosslinking reaction is controlled by a Brabender viscometer test. Typical crosslinked starches are obtained when the initial rise of the viscosity of the product is between 104 and 144 °C, and the viscosity of the product does not rise above 200 Brabender units at temperatures less than 130 °C.

The crosslinked starch slurry is then drum–dried and milled to obtain a dry product. The effectiveness of the product is checked by the API Fluid Loss Test or other standards after static aging of sample drilling fluids containing the starch at elevated temperatures. The milled dry product can then be incorporated into the oil well drilling fluid of the drill site [95, 96].

3.2.13 Pregelatinized Starch

The properties of the filter cake formed by macroscopic particles can be significantly influenced by certain organic additives. The overall mechanism of water-soluble fluid loss additives has been studied by determining the electrophoretic mobility of filter cake fines. Water-soluble fluid loss additives are divided into four types according to their different effects on the negative electrical charge density of filter cake fines [97]:

1. electrical charge density is reduced, by poly(ethylene),
2. glycol and pregelatinized starch,
3. electrical charge density is not changed by carboxymethyl starch,
4. electrical charge density is increased by a sulfonated phenolic resin, CMC, and hydrolyzed poly(acrylonitrile).

The properties of filtrate reducers are governed by their different molecular structures. Nonionic filtrate reducers work by completely blocking the filter cake pore, and anionic ones work by increasing the negative charge density of filter cakes and decreasing pore size. Anionic species cause further clay dispersion, but nonionic species do not, and both of them are beneficial to colloid stability [98].

The change of properties of the filter cake due to salinity and polymeric additives has been studied by scanning electron microscope (SEM) photography [99]. Freshwater muds with and without polymers, such as starch, PAC, and a synthetic high-temperature-stable polymer, were prepared, contaminated with electrolytes (NaCl, $CaCl_2$, $MgCl_2$), and aged at 200-350 °F (90-189 °C).

Static API filtrates before and after contamination and aging were measured. The freeze-dried API filter cakes were used for SEM studies. The filter cake structure was influenced by electrolytes, temperature, and polymers.

In an unaged, uncontaminated mud, bentonite forms a card-house structure with low porosity. Electrolyte addition increases the average filter cake pore size. Temperature causes coagulation and dehydration of clay platelets. Polymers protect bentonite from such negative effects.

3.2.14 Gellan

It has been found that gellan has good characteristics as a filtrate reducer in water-based drilling fluids [100, 101]. Preferential use is made of native gellan, which has a considerable gelling capacity and good solubility. It should be noted that native gellan contains cellular debris or other insoluble residue. Xanthan gum has been used extensively in the oil industry as a viscosifier for various applications [102]. Deacetylated xanthan gum is used in guar-free compositions instead of guar [103].

3.2.15 Humic Acid Derivates

Leonardite is a soft waxy mineraloid which can be dissolved in alkaline solutions. It is an oxidation product of lignite and has a high content of up to 90% of humic acid [104]. Leonardite is named after Arthur Gray Leonard (1865-1932), the first director of the North Dakota Geological Survey.

The humic acid may serve as a hydration buffer to help keep a powder or pellets dry and flowable [105]. In general, a hydration buffer is used to inhibit substantial absorption of water from the atmosphere.

Polysulfonated humic acid is a drilling fluid filtrate loss additive composed of three mud additives: sulfonated chromium humate, sulfonated phenolic resin, and hydrolytic ammonium poly(acrylate) [106].

The field application and the effectiveness of polysulfonated humic acid resin, especially in extra-deep wells, in sylvite and undersaturated salt muds, have been described. Polysulfonated humic acid resists high temperature, salt concentration, and calcium contamination as well. This type of drilling fluid has stable properties and good rheologic characteristics and can improve cementing quality.

The viscosity can be reduced by the addition of chrome humate compounds. This has advantages in the high-temperature stability of the drilling mud [107].

In comparison to humates as such, chrome lignites and ferrochrome lignosulfonates, chrome humates are superior with respect to a combination of good thinning properties with low shear strength and good fluid loss control. Further, the preparation of the chrome humates is easy.

Representative chromium compounds that can be used as reactants with humate to obtain chrome humates are summarized in Table 17.

The reaction proceeds as follows:

Preparation II–2: A humate suspension is diluted with water to an 18% solid concentration. This mixture is stirred and heated up to 70 °C. Then the chromium reagent, in aqueous solution, is slowly added over

Table 17 Chromium compounds [107]

Compound
Sodium dichromate dihydrate
Sodium chromate tetrahydrate
Chromic sulfate
Chromic chloride
Chromium potassium sulfate

1.5 h. Eventually, the temperature is raised to 75 °C and maintained there for 1 h. Then potassium hydroxide is added. Eventually, the total amount of water to be added is adjusted to give a final humate solids concentration of 16%. The mixture is heated for 1 h to 75-80 °C, then cooled, and freeze-dried. ∎

3.2.16 Sodium Metasilicate

Sodium silicate has been successfully used as a chemical grouting material for many years. It is used in particular during drilling of high-permeability formations [108]. When an aqueous mixture of sodium silicate and an activating agent, such as an ester, is injected into the ground, the silicate solution reacts to form a colloid, which polymerizes further to form a gel. The gel provides increased strength, stiffness, and reduced permeability in predominantly granular soils.

These properties have been utilized in water-based drilling fluid systems to prevent fluid loss while drilling high-permeability formations, particularly during drill-in and completion operations [109]. The gel produced by the silicate reaction is soluble in both acids and bases.

The upper limit on the amount of alkali metal silicate depends on the gel strength necessary and the pore size in the formation. The bigger the pore size in the formation, the higher the gel strength desired and, generally, the higher the desired concentration of alkali silicate. For practical applications, the concentration of alkali silicate generally is about 40% because most commercial silicates are available in this concentration.

Further, the drilling fluid system contains activating agents. The activating agent is effective as it hydrolyzes, thereby decreasing the pH. Example for an activating agent is formamide, or water-soluble esters. Accelerating agents may be added to accelerate the gel formation. A suitable accelerating agent is sodium aluminate [108, 109]. Examples for the formulation of drilling mud systems that use silicate for fluid loss control have been given in detail in Ref. [109].

3.2.17 Active Filter Cake

A water-based drilling fluid system which generates an active filter cake has been described in Ref. [110]. The active filter cake once formed is impermeable to an aqueous phase, thus reducing fluid loss and ensuring reduced damage to the formation, yet simultaneously is permeable to the back flow of hydrocarbons during a hydrocarbon recovery process.

Modified Starch

The fluid loss additive is selected from hydrophobically modified starch, PAC, CMC, hydrophobically modified synthetic polymers, e.g., poly(hydroxypropyl methacrylate). The most preferred fluid loss additive is a polymerized starch or a starch modified by hydroxymethylation and hydroxypropylation. As bridging material a hydrophobically coated calcium carbonate can be used [110].

Magnesium Peroxide

Magnesium peroxide is very stable in an alkaline environment and remains inactive when added to polymer-based drilling fluids, completion fluids, or workover fluids. Because the magnesium peroxide is a powdered solid, it becomes an integral part of the deposited filter cake [111]. The peroxide can be activated with a mild acid soak. This treatment produces hydrogen peroxide, which decomposes into oxygen and hydroxyl radicals (OH·) when catalyzed by a transition metal.

These highly reactive (OH·) species attack positions at the polymers that are resistant to acid alone. Significant improvements in filter cake removal can be realized by using magnesium peroxide as a breaker in alkaline water-based systems, especially in wells with a bottom hole temperature of 150 °F (65 °C) or less, in the following operations: drilling into a pay zone, underreaming, lost circulation pills, and fluid loss pills for gravel prepacks.

3.2.18 Acid Combination

The removal of filter cake generated by drilling fluids weighted with manganese tetraoxide (Mn_3O_4) particles may require strong acids [112]. Organic acids are not efficient when high-density drilling fluids are used. Here, a single stage of HCl treatment in a high-temperature and corrosive environment cannot be used. In certain cases, a two-stage treatment has been recommended to degrade first the polymers therein, and then to dissolve the solid particles in the filter cake.

An efficient technique that involves mixing HCl and an organic acid has been developed to dissolve the filter cake in a one-stage treatment [112]. A combination of 4% lactic acid and 1% HCl dissolved the Mn_3O_4 particles completely at 88 °C.

In addition, a third method was examined also being tested to remove the filter cake by using a combination of organic acid and an enzyme [112].

3.2.19 Latex

When a polymer latex is added to a water-based drilling fluid, this compound can reduce the fluid loss. The polymer latex is capable of providing a deformable latex film on the borehole wall. Latexes may include the following components [113]:

- vulcanizable groups, e.g., butadiene;
- vulcanizing agents such as sulfur, 2,2′-dithiobisbenzothiazole, organic peroxides, azo compounds, alkylthiuram disulfides, and selenium phenolic derivatives;
- vulcanization accelerators, such as fatty acids such as stearic acid, metallic oxides such as zinc oxide, aldehyde amine compounds, guanidine derivatives, and disulfide thiuram compounds;
- vulcanization retarders such as salicylic acid, sodium acetate, phthalic anhydride, and N-cyclohexyl thiophthalimide;
- defoamers;
- fillers to increase or decrease the treatment density as required.

Latex emulsions are also used in cement compositions to reduce the fluid loss [113]. Latex emulsions can be also added to reduce the brittleness of the sealant compositions and thus improve the flexibility.

Further, latex emulsions prevent the gas migration. This property is useful when the sealant starts curing, i.e., the sealant composition changes from a low viscous fluid to a highly viscous mass. During this transition phase, the sealant mass can no longer transmit the hydrostatic pressure.

When the pressure exerted on the formation by the sealant composition falls below the pressure of the gas in the formation, the gas initially migrates into and through the composition. The gas migration causes flow channels to form in the sealant composition, and those flow channels permit further migration of the gas after the sealant composition sets [113].

The latex forms a semi-permeable film or seal on the subterranean formation [114]. The polymer latex is a carboxylated styrene/butadiene copolymer.

There are commercially available carboxylated styrene/butadiene copolymers, e.g., Gencal® 7463 [115].

Sulfonated latexes have the advantage as that they can be often used in the absence of a surfactant. This may simplify the formulation and transportation of the drilling fluid additives to the production sites. A precipitating agent, such sodium aluminate, is preferably used together with the latex [115].

In the course of testing, photomicrographs showed an accumulation of latex along micro-fractures in the shale. Since the volume and the velocity of filtration flow into these cracks is very small, filtration alone cannot account for the latex accumulation at the crack throat. It seems that precipitation effects are responsible for these findings.

When sufficient latex is deposited to bridge the crack opening, the fracture is sealed and differential pressure is established across the latex. The differential pressure consolidates the latex deposit into a solid seal.

In addition, a precipitating agent is added to the composition, such as an aluminum complexing agent. Suitable aluminum complexes include sodium aluminate, aluminum hydroxide, aluminum sulfate, aluminum acetate, aluminum nitrate, or potassium aluminate [114].

Of course, a seal should be completely impermeable. However, the seal is considered to be semi-permeable, but nevertheless at least partially blocking the transmission of the fluid, which is sufficient to result in a great improvement in osmotic efficiency [115, 116].

The latex, where in general water is the continuous phase, is in turn suspended in a hydrocarbon base fluid having at least some emulsifier to suspend the polymer latex. In the best case, the polymer latex can be simply mixed with the hydrocarbon base fluid without the need for adding any more emulsifier that is usually present in such fluids. Some latex products exhibit a synergistic effect with aluminum complexes, with regard to their sealing properties. The polymer latex is a carboxylated styrene/butadiene copolymer or a sulfonated styrene/butadiene copolymer [116].

Optionally, a surfactant that behaves as an emulsifier, a wetting agent, and a weighting agent, e.g., calcium carbonate, barite, or hematite, is included. Suitable emulsifiers and wetting agents include surfactants; ionic surfactants such as fatty acids, amines, amides, and organic sulfonates; and mixtures of any of these with nonionic surfactants such as ethoxylated surfactants. The water-in-oil emulsion may consist of an oil phase, a water phase (salt or fresh), a surfactant, a weighting agent, and salts or electrolytes [117, 118].

Improved compositions for sealing subterranean zones and terminating the loss of drilling fluid, crossflows, and underground blowouts have been developed. The compositions are basically formulated from water, an aqueous rubber latex, an organophillic clay, and sodium carbonate [119].

Among the various latexes which can be utilized, those prepared by emulsion polymerization processes are preferred, e.g., a styrene/butadiene copolymer latex. The aqueous phase of the emulsion is an aqueous colloidal dispersion of the styrene/butadiene copolymer. The latex often contains small quantities of an emulsifier, polymerization catalysts, and chain modifying agents.

A latex stabilizing surfactant is ethoxylated nonyl phenol. As organophillic clay, an alkyl quaternary ammonium bentonite clay can be used. Sodium carbonate functions as buffer and prevents the destabilization of the rubber latex in the course of contact with calcium in the mixing water. As an dispersing agent, the condensation reaction product of acetone, formaldehyde, and sodium sulfite is a preferred chemical [119]. The synthesis has been described elsewhere [120, 121].

When such a composition contacts an oil-based drilling fluid, it instantly forms a resilient rubbery mass having ultra-high viscosity [119]. As the sealing mass is displaced through the wellbore, it enters and seals thief zones such as vugs and fractures through which fluid is lost. A flexible seal is obtained by a combination of extrusion pressure and friction pressure. The sealing compositions are self-diverting and plug multiple weak zones in a single well treatment.

3.2.20 Carbon Black

A drilling fluid system has been developed that combines a mixture of carbon black, asphaltite, and lignite with a mixture of fish oil and glycol [122, 123].

Carbon black is basically pure carbon and exists in extremely small particle diameters. Therefore, carbon black particles have a high surface area of 25-$500\,m^2\,g^{-1}$. They have an oil absorptive capacity of 0.45-$3\,cm^3\,g^{-1}$.

The hardness of carbon black, in addition to its high affinity for lubricating substances, makes it an excellent carrier to extremely tight fittings, such as a metal-to-metal contact. Coated with lubricant, the ultra-fine particle size penetrates openings and surfaces not normally penetrable with other solids in the drilling fluid system.

Carbon black helps improve the bacterial degradation of hydrocarbons by forming microcells which allow greater surface area exposure for the bacteria to dissipate and thus destroy the hydrocarbon. It has been determined that a hydrocarbon sheen of any significant size on the surface of water is environmentally unacceptable due to the adverse effects on marine life. Almost all drilling fluids inadvertently contain a small percentage of a sheen forming hydrocarbon. This hydrocarbon could enter the drilling fluid as the drilling assembly penetrates a hydrocarbon laden sand or by simply adding it to the drilling fluid to obtain a specific benefit.

Carbon black is strongly organophilic and has an extremely high affinity for oils, phenols, alcohols, fatty acids, and other long carbon chain products normally used in drilling fluid.

In addition, asphaltite is by nature extremely hydrophobic and will not readily mix with water or water-based drilling fluids. Thus, it is difficult to use asphaltite as an effective drilling fluid additive for water-based drilling fluids.

The addition of a nonionic surfactant may be overcome the problems with hydrophobicity. A poly(oxyethylene oxypropylene) glycol may be used or a mixture from poly(butylene glycol), poly(ethylene glycol) (PEG), and poly(propylene glycol).

The hydrophobic asphaltite is mixed with the hydrophobic carbon black and the surfactant. Then this mixture is sheared under a sufficiently high mechanical shear for a sufficient time to convert both the hydrophobic carbon black and hydrophobic asphaltite into a hydrophilic carbon black and a hydrophilic asphaltite [123].

This allows the carbon black to remain dispersed and separated into individual particles which stack or plate out on the side of the wellbore, thus to reduce fluid loss. These finely dispersed, surface coated particles act as a excellent plugging agents for improved fluid loss control [123].

3.2.21 Testing of Fluid Loss Additives

Fluid loss prevention is a key performance attribute of drilling fluids. For water-based drilling fluids, significant loss of water or fluid from the drilling fluid into the formation can cause irreversible change in the drilling fluid properties, such as density and rheology occasioning instability of the borehole. Fluid loss control is measured in the laboratory according to a standard procedure for testing drilling fluids [31].

Predictions on the effectiveness of a fluid loss additive formulation can be made on a laboratory scale by characterizing the properties of the filter cake formed by appropriate experiments. Most of the fluids containing fluid loss additives are thixotropic.

Therefore, the apparent viscosity will change when a shear stress in a vertical direction is applied, as is very normal in a circulating drilling fluid. For this reason, the results from static filtering experiments are expected to be different in comparison with dynamic experiments.

Static fluid loss measurements provide inadequate results for comparing fracturing fluid materials or for understanding the complex mechanisms of viscous fluid invasion, filter cake formation, and filter cake erosion [124].

On the other hand, dynamic fluid loss studies have inadequately addressed the development of proper laboratory methods, which has led to erroneous and conflicting results.

Results from a large-scale, high-temperature, high-pressure simulator were compared with laboratory data, and significant differences in spurt loss values were found [125].

Static experiments with pistonlike filtering can be reliable, however, to obtain information on the fluid loss behavior in certain stages of a cementation process, in particular when the slurry is at rest.

Oil-Droplet Retention

A systematic laboratory study of the flow of an oil–in–water emulsion in porous media was done to investigate the mechanisms of oil-droplet retention and their effect on the permeability [126].

Granular packs of sharp-edged silicon carbide grains and stable and dilute dodecane-in-water emulsions were used for flow experiments. In this way, the in-depth propagation of produced water was studied to find whether a residual dilute emulsion could impair the permeability during a reinjection process.

Two main mechanisms of the capture of a droplet, i.e., surface capture and straining capture were identified. The surface capture mechanism can induce significant in-depth permeability losses even at a high pore size to droplet size ratio. Further, the permeability loss is very sensitive to salinity and flow rate. In comparison to surface capture, straining capture induces more severe plugging at a lower rate of propagation [126].

3.3 Emulsifiers

3.3.1 Oleophilic Ethers

Ether carboxylic acids can be used as emulsifiers in drilling fluid systems, which contain 10-30% by water and rest of an oil phase [127].

Ether carboxylic acids can be synthesized from a fatty alcohol that is converted into an alcoholate, then reacted with an alkylene oxide, and eventually alkylated with a chloroacetic acid derivative in an alkaline medium to form the end product [128].

The ether carboxylic acids are oil-soluble and are predominantly present in the oil phase and the interfaces thereof with the water phase. The use of ether carboxylic acids leads to stable emulsions and to an improvement of the filtrate values of the drilling composition. Further, the rheology of the drilling fluid is positively influenced, even at low temperatures of lower than 10 °C [127].

3.3.2 AMPS Terpolymer

A polymer has been described that affords an improved filtration to drilling fluids even after aging at high temperatures and pressures [129]. This is a terpolymer from N-vinylcaprolactam, 2-acrylamido-2-methyl-1-propane sulfonic acid, and AAm in amounts of 15%, 20%, and 65%, respectively.

The addition of the polymer to drilling fluids or other wellbore service fluids effects an improved high-temperature and high-pressure performance and the reduction in shale erosion.

The polymer may be subjected to hydrolytic conditions in the fluid whereunder the cyclic amide ring of the N-vinylcaprolactam monomer opens to create a secondary amine structure on the backbone of the polymer. This structure results in a greater molecular volume of the polymer and also a polymer having a negative molecular surface charge [129].

3.4 Gas Hydrate Control

Gas hydrates are solid, ice-like crystals that are formed under elevated pressures and at moderately low temperatures [130]. Gas hydrates consist of water molecules which form five- and six-membered polygonal structures which combine to form a closed structure which is often called a cage. These cages totally enclose or trap a gas molecule. At high pressures, multiple cages tend to combine to form larger cages enclosing gas molecules. The resulting large crystalline assemblies are thermodynamically favored at elevated pressures. Under sufficient pressure, gas hydrates will form at temperatures well above the freezing point of water.

Primary promoters of gas hydrates are gas with free water present near its water dew point, low temperatures, and high pressures. Secondary promoters are high velocities, pressure pulsations, any type of agitation, and the introduction of a small crystal of a hydrate.

During deepwater drilling operations, all of the primary gas hydrate promoters are present. The aqueous drilling fluid supplies the free water, low temperatures are encountered on the sea floor, and the hydrostatic head of the fluid produces high pressures.

The formation of gas hydrates may become a problem in a pressure and temperature range between about 500 psia at 1.67 °C or lower and about 8000 psia at 27 °C or lower, particularly between about 1000 psia at 2 °C or lower and about 6000 psia at 27 °C or lower [130]. In summary, high pressure, low temperature, and the presence of water favor the formation of gas hydrates within a drilling fluid [131].

There are two types of inhibitors: kinetic inhibitors, which act on the kinetics of crystal growth, and thermodynamic inhibitors.

A thermodynamic inhibitor is a compound that can bind to water through intermolecular bonds, thus preventing bound water molecules from forming hydrogen bonds with each other and with the remaining free water. In other words, these compounds reduce the free water concentration, which is called water activity in the system. In this case, there emerges an additional barrier to crystal formation, and the appearance of hydrates then requires higher pressures or lower temperatures.

A preferred hydrate suppressor is ethylene glycol, which not only suppresses hydrate formation, but also improves the overall performance of the water-based drilling fluid by reducing the density of the fluid [130]. The mysid shrimp toxicity test [132] for ethylene glycol resulted in an LC_50 of 970,000 ppm SPP. This is more than 30 times the minimum EPA standard for discharge into coastal waters. Ethylene glycol has the additional advantage that it produces no sheen on the receiving waters.

Another additive formulation for water-based drilling fluids has been described. This formulation consists of a polyglycerol mixture, a carboxylic acid salt, and an inorganic salt [131]. The 20% cryoscopic depression of some of the additives used in the formulation of our invention is shown in Table 18.

R1 contains monoglycerol (12-18%), diglycerol (25-20%), triglycerol (around 5%), and other polyglycerols in a smaller proportion (less than 1%). R1 contains also sodium chloride, phosphate, and sulfate in a total proportion by weight of about 45%. R2 is a mixture of polyglycerols containing mainly monoglycerol (50-55%), diglycerol (28-32%), and triglycerol

Table 18 Cryoscopic depression [131]

Compound	Cryoscopic depression (K)
R1	7.5
R2	2.4
Sodium lactate	9.8
Sodium acetate	12.5
Sodium formate	11.8
Glycerol	5.6
Monoethylene glycol	7.8
Methanol	14.8
Sodium chloride	16.6

(10-12%). R2 also contains other polyglycerols such as tetra- or pentaglyc-erol (or higher polyglycerols) in a total percentage by weight of less than 5%.

The formulation R1 has approximately the same cryoscopic depression as monoethylene glycol. The formulation R2 has a lower cryoscopic depression than the other compounds.

The synergistic effects of the compounds in Table 18 with respect to cryoscopic depression have been studied and demonstrated [131]. In particular, a remarkable synergistic effect is obtained by combining the R2 formulation with salts. This is probably due to the interaction of the mono and polyglycerol molecules with the ionic species. These species could avoid polymerization of R2 molecules, hence greater availability of hydrophilic sites and greater lowering of the activity of water.

A series of thermodynamic experiments have been performed on 16 simulated drilling muds and associated test fluids in order to screen the equilibrium conditions for hydrate formation in water-based drilling fluids [133]. Mainly, the salt content and the glycerol content of water in mud dominate the formation of hydrates. Further, other mud additives, such as bentonite, barite, and polymers, slightly promote the hydrate formation.

3.5 Lubricants

3.5.1 Basic Studies

Extensive laboratory work has been carried out in order to determine the performance of a number of lubricants. These studies included tests to determine the potential of formation damage in several types of drilling fluids, as well as the reduction of the friction coefficient.

Certain polymer additives are effective also as lubrications as a side effect. However, in many cases, additional lubricants must be added to be successful in drilling to the total intended depth [134].

Lubricant for water-based drilling is primarily chosen due to their technical performance and due to environmental restrictions. Hydrocarbons and fatty acids were used mostly in the past, but nowadays a trend to more environmentally acceptable alternatives can be seen, in particular to esters and naturally occurring vegetable oils.

These chemicals are highly lubricious materials as they are significantly reducing the coefficients of friction in water-based fluid environments, of both metal/metal and metal/rock contacts. Laboratory tests reveal that the reduction of the coefficient of friction reaches up to 70%. Clearly, effective additives exhibit high degree of surface activity.

This property improves their adhesion to the metal casing or the drilling mud solids. In this way, the lubricity of the surfaces is enhanced. On the other hand, the surface activity makes the lubricants more prone to react with other components of the mud.

For example, lubricants may act as an emulsifier even in the presence of small quantities of oil. Thus, such a composition may turn basically into an invert emulsion with the consistency of cottage cheese [134]. Of course, such events are highly undesirable as the formation of a highly viscous material is definitely a drilling hazard. Moreover, in the worst case the production zone may be damaged.

Apart from this, the lubricant may react with and divalent or multivalent ions, forming ionic bonds as found in ionomers. This reaction results in the formation of a grease like precipitate. The precipitate may formed already at low concentrations of calcium or magnesium ions of 1000 ppm, depending on the chemical nature of the lubricant. Such ionic concentrations are frequently observed in even fresh water. All these issues must be taken into account for the selection of proper lubricants for water-based drilling fluids [134].

3.5.2 Biodegradable Olefin Isomers

Certain olefin isomers can be added to WBMs as downhole lubricants [135]. Synthetic poly(α-olefin)s are nontoxic and effective in marine environments when used as lubricants. The olefin isomers may be formed by polymerizing ethylene. Ethylene is generally derived from the catalytic cracking of naphtha.

3.5.3 Fatty Acid Esters

Certain esters improve the lubricating properties of water-based drilling muds. Oligoglyercol fatty acid ester additives serve in particular to improve the lubricating action of water-based drilling muds [136]. The oligoglycerol esters can be obtained by the acid-catalyzed or base-catalyzed esterification directly from oligoglycerol with the corresponding fatty acids.

Oligoglycerol esters are liquid at room temperature and are a mixture of various oligoglycerol esters. The esters accordingly contain mixtures of di- and triglycerol and tetra- and pentaglycerols and glycerol [136]. The fatty acids are given in Table 19 and some fatty acids are shown in Figure 9.

Table 19 Fatty acids [136]

Acid compound	IUPAC name
Caproic acid	Hexanoic acid
Caprylic acid	Octanoic acid
2-Ethylhexanoic acid	2-Ethylhexanoic acid
Capric acid	Decanoic acid
Lauric acid	Dodecanoic acid
Isotridecanoic acid	11-Methyldodecanoic acid
Myristic acid	Tetradecanoic acid
Palmitic acid	Hexadecanoic acid
Palmitoleic acid	*cis*-Hexadec-9-enoic acid
Stearic acid	Octadecanoic acid
Isostearic acid	16-Methylheptadecanoic acid
Oleic acid	*cis*-9-Octadecenoic acid
Elaidic acid	*trans*-9-Octadecenoic acid
Petroselinic acid	*cis*-6-Octadecenoic acid
Linoleic acid	(9*Z*,12*Z*)-9,12-Octadecadienoic acid
Linolenic acid	(9*Z*,12*Z*,15*Z*)-9,12,15-Octadecatrienoic acid
Elaeostearic acid	(9*Z*,11*E*,13*E*)-9,11,13-Octadecatrienoic acid
Arachidic acid	Icosanoic acid
Gadoleic acid	*cis*-9-Icosenoic acid
Behenic acid	Docosanoic acid
Erucic acid	*cis*-Docos-13-enoic acid

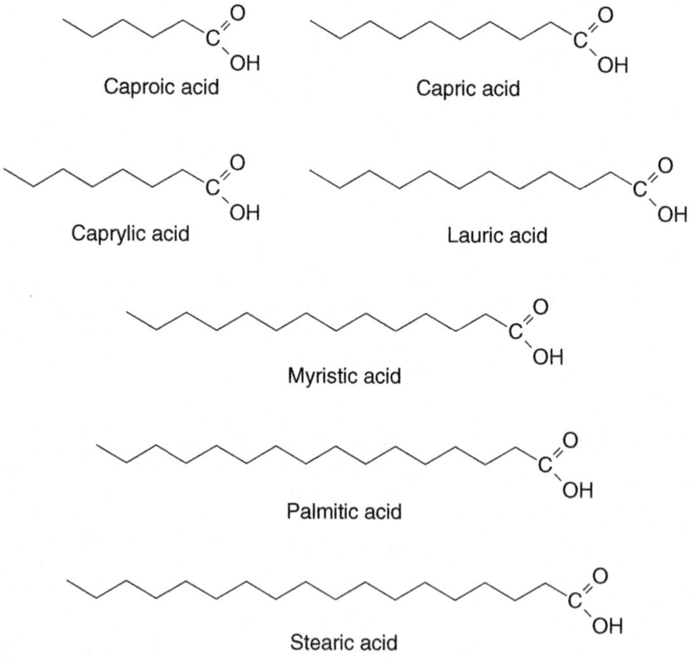

Caproic acid   Capric acid

Caprylic acid   Lauric acid

Myristic acid

Palmitic acid

Stearic acid

Figure 9 Fatty acids.

Biodegradable Esters

There is a growing interest in alternatives with better biodegradability, in particular esters. Some esters are summarized in Table 20.

The use of esters in water-based systems particularly under highly alkaline conditions can lead to considerable difficulties. Ester cleavage can result in the formation of components with a marked tendency to foam which then introduce unwanted problems into the fluid systems.

Sulfonates of vegetable oils, in particular soya oil sulfonate, are also used as lubricants in practice. Soya oil sulfonate can be used in water- and oil-based systems, but shows significant foaming, especially in water-based fluids, which restricts its usefulness [138].

A lubricating composition that comprises components that can be obtained from byproducts of manufacturing processes that provide a use for such byproducts other than disposal would also be a significant contribution to the art and to the economy [139].

The lubricating composition has been proposed that consists of esters obtained by the reaction product of a glycerol component comprising glycerol, and glycerol oligomers, and a fatty acid component. The reaction product is neutralized with potassium hydroxide or ammonium hydroxide. The composition of the glycerol component is shown in Table 21.

The fatty acid component can be obtained from vegetable oils, from wood pulp processing, from animal fats processing, etc. Examples of catalysts for esterification include sulfuric acid, hydrochloric acid, nitric acid, p-toluene sulfonic acid, etc. A preferred catalyst is concentrated sulfuric acid [139].

In addition, the composition can be subjected to chain extension using diacids, such as maleic acid, succinic acid, or glutaric acid [139].

Ester-Based Oils

Several ester-based oils are suitable as lubricants [140, 141], as are branched chain carboxylic esters [142]. Tall oils can be transesterified with

Table 20 Esters as lubricants

Ester compound	Reference
2-Ethylhexyl oleate	[137]
Triglyceride oil	[137]
Soya oil sulfonate	[138]
Glycerol monotalloate	[138]
Sulfonated castor oil	[138]

Table 21 Composition of the
glycerol component [139]

Compound	%
Glycerol	10-13
Diglycerol	16-23
Triglycerol	5-7
Tetraglycerol	4-6
Pentaglycerol	3-4
Heavier polyglycerols	15
NaCl	2-4
Na_2CO_3	0.3-1
Water	22-28
Carboxylic acid salt	11-14

glycols [143] or condensed with monoethanol amine [144]. The ester class also comprises natural oils, such as vegetable oil [145], spent sunflower oil [146–149], and natural fats, e.g., sulfonated fish fat [150]. In WBM systems, no harmful foams are formed from partially hydrolyzed glycerides of predominantly unsaturated C_{16} to C_{24} fatty acids.

The partial glycerides can be used at low temperatures and are biodegradable and nontoxic [151]. A composition for high-temperature applications is available [152]. It is a mixture of long-chain poly(ester)s and PAs.

In the case of esters from, e.g., neopentylglycol, pentaerythrite, and trimethylolpropane with fatty acids, tertiary amines, such as triethanol amine, together with a mixture of fatty acids, improve the lubricating efficiency [153].

Ester Alcohol Mixtures

In addition to esters, mixtures of fatty alcohols with carboxylic acid esters have been proposed as a lubricating additive in water-based drilling fluids. The alcohols include guerbet alcohol and oleyl alcohol with oleyl oleate or isotridecyl stearate as ester component [154]. Fatty alcohols exhibit a foam-suppressing effect.

Phosphite Esters

Phosphites are useful as lubricating aids. A suitable phosphite ester is oleyl phosphite. Dihydrocarbyl phosphites can be prepared from a lower molecular weight dialkylphosphite, such as dimethyldialkylphosphite.

This compound is reacted with longer chain alcohols, either linear or branched [155].

For example, a method of preparation is as:

Preparation II–3: A mixture of 2-ethylhexanol, Alfol 8-10, and dimethylphosphite is heated to 125 °C while sparging with nitrogen. The emerging methanol is removed by distillation. After 6 h, the mixture is heated to 145 °C and kept at this temperature for additional 6 h. The reaction mixture is then stripped to 150 °C at 50 mm Hg where additional distillate is recovered. The residue is then filtered to get the desired mixed dialkyl hydrogen phosphite. ∎

Phosphate Esters

It has been found that the inclusion of poly(ether) phosphate esters in combination with PEG can give aqueous drilling fluids which provide good lubricating properties in a wide range of drilling fluids [156].

Typical synthetic routes to such esters involve the reaction of the poly(ether) with a phosphating agent such as phosphorus pentoxide or polyphosphoric acid. The use of polyphosphoric acid in the synthesis gives higher proportions of the monoester, which is preferred. The optimal molecular weight of the PEG is 400 D [156]. The compatibility of the lubricant may be adversely affected by other components of the drilling fluid, particularly by divalent cations such as calcium [156, 157].

Organic Sulfides

Hydroxy thioethers contribute to oxidative stability and anti-wear properties. For oil-in-water emulsions such a friction modifier can be incorporated into the oil phase [158]. n-Dodecyl-(2-hydroxyethyl) sulfide may be prepared by condensing 1-dodecene with mercaptoethanol. The corresponding bis ether, i.e., 2,2′-di-(n-dodecylthio)-diethyl ether can be prepared by condensing the alcohol in the presence of an acid catalyst, such as sulfuric acid or methane sulfonic acid.

3.5.4 Silicate-Based Compositions

During drilling operations, sometimes gumbo shale formations are encountered. Such formations comprise highly reactive shale clays, lose physical, and lose dimensional integrity when they are exposed to the pure water or the water filtration of water-based drilling fluids [159].

Although oil-based drilling fluids can effectively handle reactive shale clays, these fluid types pose the problems of safety, environmental hazards, and interference with oil and gas well logging operations.

In order to suppress the swelling and dispersing of the shale clays in water-based drilling fluids, a variety of special water-based drilling fluids have been developed and used.

The use of silicate as a component in drilling fluids is well established. Silicate has been used since the late 1930s [160].

Until recently, silicate did not achieve widespread use, because of certain advantages of oil-based drilling fluids. Namely, oil-based drilling fluids offer ease of use, are not prone to gellation or precipitation, and provide a good lubricity between drill string and wellbore.

However, in the recent past, environmental pressures have caused strength to improve the performance deficiencies in silicate drilling fluids. So, a suitable replacement for oil-based drilling fluids in some of the more difficult fields became a goal of the drilling industry [108, 161].

One of the most inhibitive WBMs in commercial use is based on silicates. It is typified by M-I Swaco's SILDRIL. In terms of inhibition, these muds are not as effective as OBMs, but are significantly better than other WBMs, including glycols. Instead, silicate muds may suffer from drawbacks, including health and safety concerns, due to their high pH, poor thermal stability and lubricity, intolerance to contamination, high maintenance costs, detrimental effects on some downhole equipment, and potential for causing formation damage [162].

Potassium silicate-based drilling fluids have become a more preferred inhibitive water-based drilling fluid for the drilling in reactive shale formations. Nowadays, potassium silicate drilling fluids provide a very favorable wellbore stability for shale sequences, where traditionally only oil-based drilling fluids were successful [159].

However, potassium silicate drilling fluids experience high coefficients of friction between drill string and wellbore, owing to the silicate coating on the surfaces, making the surfaces of drill string and wellbore rough.

This drawback can be improved by the addition of lubricants. A lubricant composition has been developed that is a synergistic combination of extreme pressure lubricants.

Such lubricants may be lecithins, sulfurized vegetable oil, sulfurized lard oil, sulfurized vegetable esters or sulfurized lard esters, and modified castor oil [159]. Lubricant compositions are summarized in Table 22.

Table 22 Lubricant compositions [159]
Formulation name

Mixture of sulfurized vegetable oil with 5% sulfur content and liquid paraffin carrier oil	KT 810
Mixture of sulfonated castor oil with 0.5% sulfonation degree and liquid paraffin carrier oil	KT 811
Mixture of sulfurized vegetable oil with 2% sulfur content and liquid paraffin carrier oil	KT 812
Mixture of lecithin and liquid paraffin carrier oil	KT 813
Mixture of sulfurized vegetable oil with 5% sulfur content, sulfonated castor oil with 0.5% sulfonation degree and liquid paraffin carrier oil	KT 814
Mixture of sulfurized vegetable oil with 5% sulfur content, lecithin and liquid paraffin carrier oil	KT 815
Mixture of sulfurized vegetable oil with 5% sulfur content, lecithin, sulfonated castor oil with 0.5% sulfonation degree and liquid paraffin carrier oil	KT 816
Mixture of sulfurized vegetable oil with 2% sulfur content, lecithin and liquid paraffin carrier oil	KT 817
Mixture of sulfonated castor oil with 0.5% sulfonation degree and liquid paraffin carrier oil	KT 818

Additive [a]	Friction coefficient	Torque reduction (%)
–	0.39	
KT 810	0.33	15.4
KT 811	0.36	7.7
KT 812	0.34	12.8
KT 813	0.27	30.8
KT 814	0.31	20.5
KT 815	0.23	41.0
KT 816	0.19	51.3
KT 816	0.16	59.0
KT 817	0.28	28.2
KT 818	0.37	5.1

[a] $20 \, l \, m^{-3}$ added to a controlled silicate drilling fluid.

3.5.5 Sulfonated Asphalt

Asphalt is a solid, black-brown to black bitumen fraction, which softens when heated and re-hardens upon cooling. Asphalt is not water soluble and difficult to disperse or emulsify in water.

Sulfonated asphalt can be obtained by reacting asphalt with sulfuric acid and sulfur trioxide. By neutralization with alkali hydroxides, such as

NaOH or NH_3, sulfonate salts are formed. Only a limited portion of the sulfonated product can be extracted with hot water. However, the fraction thus obtained, which is water soluble, is crucial for the quality.

Sulfonated asphalt is predominantly utilized for water-based drilling fluids but also for those based on oil [163]. Apart from reduced filtrate loss and improved filter cake properties, good lubrication of the drill bit and decreased formation damage are important features assigned to sulfonated asphalt as a drilling fluid additive [163].

In particular, clay inhibition is enhanced by sulfonated asphalt in the case of water-based drilling fluids. If swellable clays are not inhibited, undesirable water absorption and swelling of the clay occurs, which can cause serious technical problems including the instability of the borehole, or even a stuck pipe.

The mechanism of action of sulfonated asphalt as a clay inhibitor in a drilling fluid is explained that the electronegative sulfonated macromolecules attach to the electropositive ends of the clay platelets. Thereby, a neutralization barrier is created, which suppresses the absorption of water into the clay.

In addition, because the sulfonated asphalt is partially lipophilic, and therefore water repellent, the water influx into the clay is restricted by purely physical principles. As mentioned already, the solubility in water of the sulfonated asphalt is crucial for proper application. By the introduction of a water soluble and an anionic polymer component, the proportion of water–insoluble asphalt can be markedly reduced.

In other words, the proportion of the water soluble fraction is increased by introducing the polymer component. Especially suitable are lignosulfonates as well as sulfonated phenol, ketone, naphthalene, acetone, and amino plasticizing resins [163].

3.5.6 Graphite

Graphite has been used for many years as lubricant [164]. The lubricating mechanism of graphite is thought to be mechanical in nature and results from the sliding of one graphite particle over another graphite particle.

Graphite may be used as a dry lubricant or may be dispersed in a lubricating oil. Graphite particles may be also incorporated into a grease product for improved lubrication.

Because graphite has the reputation for being a superior lubricant, various grades of graphite have been tested in various water-based drilling fluids [164]. It was concluded that dry graphite added to a water-based drilling fluid reduced the friction only very little. Since all the tests using dry graphite proved unsuccessful, it was concluded that the surface of the graphite was hydrophobic or organophilic.

The lubricating qualities of graphite in a WBM can be improved by coating the surface with a glycol carrier. In this way, the surface of the graphite particles becomes hydrophilic [164]. A suspension of graphite in glycol with an extended shelf life has been formulated. Additional solids such as uintaite improve the suspension. The hydrophilic glycol-coated graphite particles should be ideal particle plugging agents and it is believed that the graphite particles or platelets would stack one on top of the other.

3.5.7 Paraffins
Nontoxic, biodegradable paraffins have been reported, which may be used as lubricants, rate of penetration (ROP) enhancers, or spotting fluids for WBMs [165]. These may be cycloparaffins with 8–16 carbon atoms.

Purified paraffins may be produced from refined petroleum and purified by treatment with chemicals to remove sulfur and nitrogen containing impurities, as well as unsaturated hydrocarbons. Purified paraffins contain little or no detectable levels of aromatics, which are removed from refined paraffins by hydrogenation [165].

3.5.8 Partial Glycerides
Historically, the class of pure water-based systems is the oldest in the development of drilling fluids. However, their use is involved with attended by such serious disadvantages that, hitherto, only limited application has been possible for technically demanding drilling operations. Above all, the interaction of the water-based drilling fluids with the water-sensitive layers of rock more particularly corresponding layers of clay to be drilled leads to unacceptable interference with the drilling process [138].

Even in highly sensitive shale formations, adequate stability can be obtained in the case of purely water-based drilling fluids. This observation involves the use of systems based on soluble alkali metal silicates, i.e., water-glasses. However, using water-based drilling fluids requires the addition of lubricants. These include mineral oils, animal and vegetable oils and esters. The increasingly stricter regulations with regard to the biodegradability of drilling fluids and their constituents are gradually restricting the use of the otherwise particularly suitable mineral oils.

Lubricants for both WBMs and OBMs for use at low temperatures have been found as fatty acid partial glycerides. Basic WBMs and OBMs that are used for testing the lubricants are given in Table 23.

The effectiveness of the lubricants can be measured by the Almen–Wieland test [166], the Falex pin and vee block method [167], the Timken wear and lubricant test [168], and the four-ball test [169]. The effects of various lubricants have been measured as shown in Figure 10.

Table 23 WBMs and OBMs [138]

Water based		Oil based	
Water	4 l	Mineral oil	675 ml
Xanthan gum	20 g	Water	225 ml
Bentonite	56 g	CaCl$_2$	95 g
CMC	40 g	Emulsifier	35 g
Barite	1.8g	Fluid loss additive	10 g
		Viscosifier	25 g
		Lime	17 g
		Barite	360 g

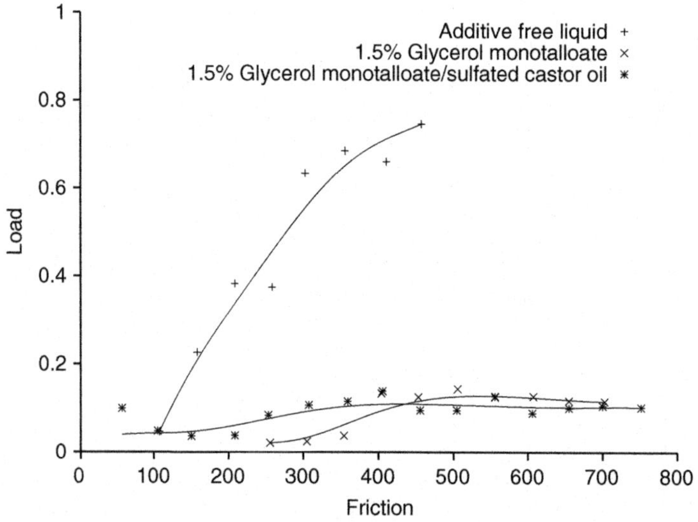

Figure 10 Effect of various lubricants (arbitrary units) [138].

3.5.9 Aminoethanols

High pH values are difficult conditions for the stability of lubricating products, in particular those based on conventional esters which hydrolyze at a high pH value and under the effect of the temperature.

Instead of alcohols, amino alcohols can be used. For example, a lubricating composition has been synthesized by the reaction of polymerized linseed oil with diethanol amine, at 160 °C. A product with a viscosity of around 2700 mPa s at 40 °C is obtained [170]. The viscosity can be reduced by adding some methyl oleate to the reaction product.

When added to a silicate mud, good lubricating properties are obtained, even at high pH, up to 12. Tests using a lubricity tester revealed that the addition of 3% of a lubricant to a base mud reduces the torque readings by 50%.

3.5.10 Polymeric Alcohols

Synthetic poly-α-olefins are nontoxic and effective in marine environments when used as lubricants, return-of-permeability enhancers, or spotting fluid additives for WBM. A continuing need exists for other nontoxic additives for WBM, which serve as lubricants, return-of-permeability enhancers, and spotting fluids.

Both poly(alkylene glycol) [171] and side chain polymeric alcohols such as PVA have been suggested. These substances are comparatively environmentally safe [172, 173].

PVAs may be applied as such or in crosslinked form [174]. Crosslinking agents can be aldehydes, e.g., formaldehyde, acetaldehyde, glyoxal, and glutaraldehyde, to form acetals, maleic acid or oxalic acid to form crosslinked ester bridges, or dimethylurea, poly(acrolein), diisocyanate, and divinylsulfonate [175].

An amine-terminated poly/oxyalkylene having an average molecular weight from about 600-10,000 D can be acylated with a succinic acylating agent, e.g., hexadecenyl succinic anhydride or a Diels-Alder diacid, obtained from an unsaturated fatty acid [176, 177]. Similarly, alkyl-aryl sulfonate salts can be used in lubrication [178].

The pendant hydroxy groups of ethylene oxide (EO)-propylene oxide (PO) copolymers of dihydroxy and trihydroxy alcohols may be sulfurized to obtain a sulfurized alcohol additive. This is effective as a lubricant in combination with oils and fats [179, 180]. The sulfurized alcohols may be obtained by the reaction of sulfur with an unsaturated alcohol. Furthermore, fatty alcohols and their mixtures with carboxylic acid esters as lubricant components have been proposed [181].

3.5.11 Fatty Acid Polyamine Salts

WBM lubricants using a blend of fatty acid polyamine salts and fatty acid esters result in a synergistically better lubricity as when either component is used separately [182].

Blends with different ratios of fatty acid diethylenetriamine salt and fatty acid methyl ester demonstrate a much better lubricity in water-based drilling fluids than those, where only fatty acid diethylenetriamine salt or fatty acid methyl ester are separately used [182].

3.5.12 Powder-form Lubricants

When a lubricant additive can be blended in powder form the handling becomes easier. Powder-form lubricant additives for water-based drilling fluids contain highly disperse silicas and fatty acid alkyl esters [183].

Highly disperse silicas are precipitated silicas obtained by the wet chemical route from alkali metal silicate solutions by the reaction with mineral acids. Other highly disperse silicas are pyrogenic silicas and pyrogenic hydrophobicized silicas. Pyrogenic silicas are obtained by the coagulation from the gas phase at high temperatures. This is normally effected by flame or high-temperature pyrolysis. The fatty acids are cocosalkyl or tallow alkyl fatty acids [183].

The resulting formulations are used in the form of a fine, free flowing powder, which is easy to dose in a temperature range of -40 to $+80\,°C$.

The quality of the lubricating properties with the above described lubricant additives was tested with a Reichert friction wear balance [184, 185]. There, a tightly fixed test roller is pressed by a double lever system onto a rotating abrasion ring where the lower third is immersed in the test liquid [183].

3.5.13 Multiphase Lubricant Concentrates

Multiphase lubricant concentrates as an additive for water-based drilling fluids and borehole servicing fluids have been described in Ref. [186].

The oil phase can be an ester mixture of substantially saturated fatty acids based on palm kernel oil and 2-ethylhexanol, hydrogenated castor oil, or glycerol monolaurate.

Heating of the multiphase formulations to temperatures in the phase inversion temperature range or higher followed by cooling to temperatures below the phase inversion temperature leads to water-based oil-in-water emulsions with an extremely fine-particle oil phase. The particle fineness of the emulsions can be so great that their particles are no longer optically visible. Instead, the emulsions have a transparent, opalescent appearance.

This state can be preserved over relatively long periods of storage by adequately lowering the temperature of the mixture below the phase inversion temperature range [186].

3.5.14 Filming Amines

A downhole well lubricant that has the additional properties of coating metal surfaces submerged in a water-based well liquid has been described in Ref. [187]. The lubricant is made from a soap, a filming amine, and an activator. Furthermore, a viscosifier and a diluent can be used.

Filming amines are widely used as corrosion inhibitors. The filming amine has a specific affinity to metal and thus effectively coats the metal equipment in the well. When added to a water-based drilling or completion fluid, the lubricant disperses and shortly coats the metal surfaces in the well.

For example, when the outside surface of a coiled tubing string rubs against the inside surface of a production string, a greasy lubricious emulsion forms on both of the surfaces. The longer the surfaces are rubbed together, the more emulsion is created. The emulsion lubricates the coiled tubing and the inside of the production string, thereby reducing the drag [187].

3.6 Density Control

3.6.1 Ionic Liquids

A drilling fluid for treating a borehole, which may have a high density while having a low amount of weighting agents, is based on a ionic liquid additive [188]. This effects a decreased tendency to separate or to precipitate the inert weighting substances from the drilling fluid.

The ionic liquid may comprise a single ionic liquid, or a mixture of different ionic liquids. Ionic liquids are melts of low-melting salts with melting points equal to or below $100\,°C$ [189]. The melting temperature of $100°C$ is chosen arbitrarily by definition.

The use of a drilling fluid with an ionic liquid may increase the density of the drilling fluid, since ionic liquids may have a higher density than water and oil, which are the known base materials for drilling fluids.

Thus, the use of a drilling fluid based on a ionic liquid may enable to design a drilling fluid which needs no weighting agents in order to achieve a suitable density.

Thus, it may also be possible to omit further additives, like polymeric additives, which are used in water- and oil-based drilling fluids to increase the amount of weighting agents that can be used in the drilling fluid. Additionally, by omitting these additives may also increase the temperature stability since ionic liquids tend to decrease the temperature stability. Another possible advantage is that the solubility of gases in the drilling fluid could be reduced. Actually, in some formulation, water can be omitted [188].

3.7 Shale Stabilization

3.7.1 Shale-Erosion Behavior

Numerous shale-stability issues can occur while drilling with WBMs, including shale sloughing and cutting disintegration [190]. These issues can be detrimental to the formation and pose difficulties with respect to rheology control, possibly reducing the ROP. The shale-erosion test is a well-known laboratory test used to characterize the erosion of cuttings in WBMs.

A mathematical modeling tool known as an artificial neural network (ANN) used to model the erosion behavior of shale cutting in WBM.

This model establishes complex relationships between a set of inputs and an output based on computational modeling. For ANN modeling of the shale-erosion behavior, the shale mineralogy and fluid composition constitute a set of inputs, while experimentally obtained the percentage erosion or the percentage of recovery of the cuttings from the shale-erosion test represent the output.

Experimental data for building the ANN model could be obtained by performing some 150 standard shale-erosion tests using five different shales with varying mineralogy and WBMs with varying salt concentrations and types, shale stabilizers, and mud weights.

The ANN model could be successfully validated for an independent set of shale-fluid interactions. The erosion behavior of the cuttings could be predicted, thereby reducing the number of trials necessary in technical service laboratories.

The mud engineers can use this model on a real-time basis as the shale chemistry varies with the depth of the formation drilling. The model could provide a convenient measurement of fluid performance, enabling fluid optimization necessary to obtain desired shale behavior in advance, thereby minimizing drilling risks and costs associated with these oftentimes unpredictable shales [190].

3.7.2 Alkyl Ethoxylates

Water-soluble glycols or polyols are widely used as chemical additives for improved shale inhibition in WBMs.

Table 24 Surfactants [162]

Trade name	Group	n	M_w (D)	HLB number
BRIJ 72	Stearyl (C18)	2	358	4.9
BRIJ 76	Stearyl (C18)	10	710	12.4
BRIJ 78	Stearyl (C18)	20	1150	15.3
BRIJ 721	Stearyl (C18)	21	1194	15.5
BRIJ 58	Cetyl (C16)	20	1122	15.7
BRIJ 98	Oleyl (C18_1)	20	1148	15.3

Alkyl ethoxylate additives for shale control have been proposed.

Aqueous solutions containing an individual alkyl ethoxylate surfactant having the formula $RO(CH_2CH_2O)_nH$, or a blend of such surfactants have been prepared from BRIJ® series from ICI [162]. Some of these surfactants are detailed in Table 24.

3.7.3 Poly(acrylamide)s

High molecular weight polyacrylamide polymers are commercially available and well known for their ability to impart borehole stability by inhibiting shale hydration.

Formulations with enhanced performance have been reported [191]. This is a drilling fluid composition that contains a nonionic low molecular weight polyacrylamide, a nonionic high molecular weight polyacrylamide, long-chain alcohols or polyols, and PAC.

The fluid forms a water blocking barrier preventing hydration of water-sensitive formations such as shales during drilling. It is not effective as a lost circulation fluid, a fluid loss additive, or as a fracturing fluid. An advantage of the fluid is that a high rate of return permeability of the formation is observed. Return permeability rates of about 86% or higher are typical with such fluid formulations [191].

A quadripolymer of AAm, 2-acrylamido-2-methyl-1-propane sulfonic acid, N-vinylpyrrolidone, and dimethyl diallyl ammonium chloride has been synthesized by solution free radical polymerization [192].

The rheological and filtrate properties of saturated brine-based fluid with the quadripolymer have been investigated before and after thermal aging tests. The apparent viscosity, plastic viscosity, and the yield point increased and the filtrate volume decreased with the increase of the concentration of the polymer after a thermal aging test at 180 °C for 16 h.

The quadripolymer could better improve the quality of a filter cake in comparison with a terpolymer composed from AAm, 2-acrylamido-2-methyl-1-propane sulfonic acid, and N-vinylpyrrolidone. In addition, the quadripolymer exhibits an excellent tolerance to salt and high temperature [192].

3.7.4 Quaternized Amine Derivatives

A water-based drilling fluid has been described that contains an aqueous-based continuous phase, a weighting material, a shale hydration inhibition agent, and a shale encapsulator in Ref. [193].

The shale encapsulator is a copolymer of amine-based monomers and vinyl acetate. In this way, quaternized amine derivatives of poly(vinyl alcohol) (PVA) are formed. Suitable examples of anions that are useful include halogen, sulfate, nitrate, formate, etc. [194].

By varying the molecular weight and the degree of amination, a wide variety of products can be tailored. In addition, conventional additives may be included into the formulation of the drilling fluid, including fluid loss control agents, alkaline reserve and other pH control agents, bridging agents, lubricants, anti-bit balling agents, corrosion inhibition agents, surfactants, and solids suspending agents [193]. It is possible to create shale encapsulators for the use in low salinity, including freshwater [194]. The repeating units of quaternized etherified Poly(vinyl alcohol) and quaternized PAM are shown in Figure 11.

Another method is to react PVA with acrylonitrile and this product is then hydrogenated to produce an amine-based PVA [193].

3.7.5 Polymeric Nonionic Amines

Poly(amine) water-based drilling fluids have been used around the world in recent years [195].

A poly(amine) shale hydration inhibitor (SDJA-1) has been closely investigated. This compound can suppress the clay hydration effectively and provides a pH buffer effect.

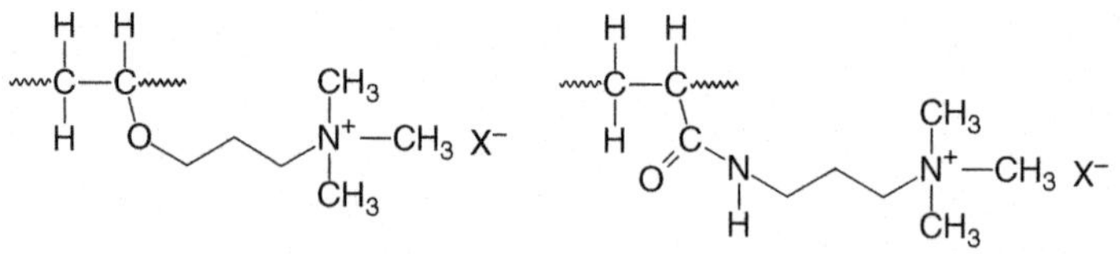

Figure 11 Quaternized etherified poly(vinyl alcohol) and quaternized PAM [194].

Also, inhibition, lubrication, and toxicity were evaluated and compared with other inhibitive water-based drilling fluids. The system is nontoxic and environmental friendly and superior to oil-based drilling fluids [195].

Many conventional shale hydration inhibition agents are of cationic type [196]. For example, triethylamine diethyl sulfate and triethylamine dimethyl sulfate provide good shale inhibiting characteristics as well as good toxicity results [197].

The cationic character may exchange cations from the surface of the shale. While such mechanism for shale hydration could be suitable for offshore wells, in land-based drilling there is a need for fluids with a low electrical conductivity. In particular, a low electrical conductivity is desired for the proper disposal of the cuttings.

A fluid with a low electrical conductivity is a fluid with an electrical conductivity of no more than $10,000 \, \mu S \, cm^{-1}$.

When polymeric nonionic amines are used, a shale hydration inhibition may be achieved without increasing the electrical conductivity of the fluid.

Such amines can be formed by the condensation of tertiary trihydroxy alkyl amines, such as trimethanol amine, triethanol amine, and tripropanol amine. The basic structure of such polymeric hydroxy amines is shown in Figure 12. A fluid composition using a polymeric amine is listed in Table 25.

Poly(ether diamine) has been extensively used as a shale hydration inhibitor with a remarkable performance [198]. A series of evaluation

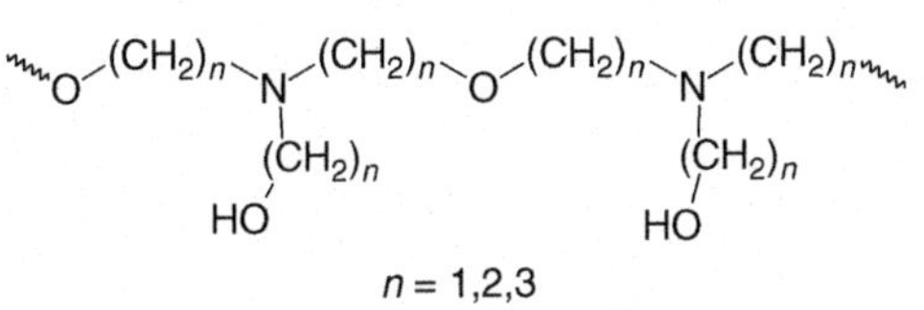

$n = 1,2,3$

Figure 12 Polymeric trihydroxy amines [196].

Table 25 Low conductivity fluid composition [196]

Compound	Example	Amount (ppb)
H_2O	Freshwater	309
Scleroglucan viscosifier	BIOVIS®	1.67
Starch derivative	FLOTROL®	3
KOH	Potassium hydroxide	drops
Nonionic polymer	EMI-1994	1
Polymeric nonionic amine	EMI-1993	5
$BaSO_4$	Barite	65

methods, including the bentonite inhibition test and a tertiary cutting rolled dispersion test, were done in order to compare the inhibition diversity between (2,3-epoxypropyl)trimethylammonium chloride and the poly(ether diamine). Also, the compatibility with bentonite and several anionic and amphoteric additives was investigated. The studies show that poly(ether diamine) exhibits a superior clay stabilization in comparison with (2,3-epoxypropyl)trimethylammonium chloride and provides a more permanent effect. In contrast to (2,3-epoxypropyl)trimethylammonium chloride, the poly(ether diamine) is compatible with most of the conventional additives and bentonite. This alleviates the compatibility problems of cationic fluids. In addition, the poly(ether amine) is environmentally accepted and has been evaluated by a toxicity test.

3.7.6 Poly(propyleneimine) Dendrimers

Shale hydration inhibition can be made from hydrogenated poly (propyleneimine) dendrimers, or polyamine twin dendrimers. Under favorable conditions, these compounds are not hydrolyzed at temperatures of 40-260 °C.

The synthesis is shown in Figure 13.

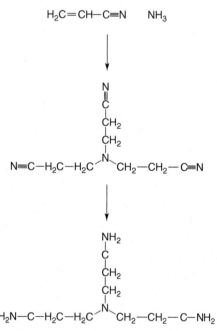

Figure 13 Dendrimers [199].

3.7.7 Copolymer of Styrene and Maleic Anhydride

An additive for shale stabilization has been described for water-based fluids [200]. This is a copolymer of styrene and maleic anhydride with side chains from EO or PO.

Also, partially hydrolyzed polymeric vinyl acetate homopolymers and copolymers have been found to be suitable [201]. The presence of AA as the comonomer inhibits crystallization and thereby promotes the water solubility.

Alkylene oxide-based side chains can be introduced into the polymer by the esterification of the carboxylic or anhydride groups with a polyoxyalkylated alcohol. The esterification step can be carried out either before or after the polymerization reaction. Preferably some 50% of the functional carboxyl carbon atoms in the polymer should be esterified [200]. The synthesis of such grafted polymers has been described in detail in Ref. [202].

3.7.8 Anions with High Hydrodynamic Radius

In the drilling of oil wells with water-based fluids often instability problems are found in the shale [203]. These problems arise due to two main factors: The reactivity of claystone to water and the propagation of the pressure from the mud column into the formation, which modifies the stress within the rock causing it to break.

A composition has been reported, which is particularly effective in reducing the pressure of drilling fluids by an osmotic effect [203]. This composition contains anions with a high hydrodynamic radius. Suitable compounds a high hydrodynamic radius are shown in Table 26 and in Figure 14.

Actually, picric acid is not an acid in the chemical sense. However, the aromatic structure and the pendent nitro compounds allow the hydroxy group to become to behave like a carboxyl group, as the proton dissociates easily.

Table 26 Anions with high hydrodynamic radius [203]

Anionic compound	Anionic compound	Anionic compound
Dimethylmalonate	3,5-Dinitrobenzoate	Crotonate
Fumarate	Glutarate	Malate
Naphthylacetate	Picrate	Pivalate

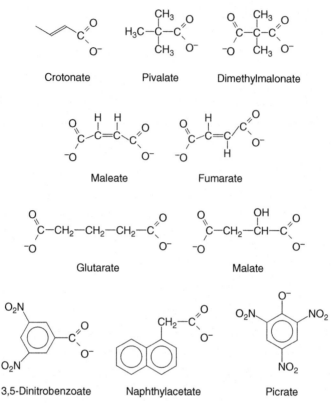

Figure 14 Anions with high hydrodynamic radius.

Pressure Transmission Test

The pressure transmission test used to verify the osmotic effect. So it is possible to measure the pressure of the formation fluids and drilling fluid. The greater the difference in pressure between the mud and the formation fluids, the more effective is the formulation. For example, potassium pivalate has been found to be highly effective for such purposes [203].

Subsequently, the pressure transmission test is described in Ref. [204]: A sample of claystone is inserted in a rubber sheath, closed at the ends by two steel heads, and is placed in the cell, where it is pressurized and heated. The two steel heads are connected to two independent hydraulic circuits, which operate under high-pressure conditions. One is an open circuit where a solution of pore water or a drilling fluid is pumped and laps against one of the two sides of the cylindrical sample. The pressure of the mud is kept constant using a back pressure regulator. The volume of the part of the circuit under pressure during the test is $22 \, cm^3$. Furthermore, a piezoresistive transducer measures continuously the internal pressure of the circuit, just before the fluid enters the cell.

The other circuit is the pore circuit, which simulates the shale, has an internal volume of $22 \, cm^3$, with an additional tank of $38 \, cm^3$, so a total volume of $60 \, cm^3$. This tank contains a fluid with a composition similar to the pore fluid, whose pressure is measured by a piezoresistive transducer. A computer records the cell temperature, the pressure of the pore circuit and of the mud circuit. The experiments are carried out according with respect to consolidation and evaluation of the drilling fluid.

Consolidation

The sample of claystone is brought to the desired temperature and pressure conditions. For example, this is maintained for 2 d at a temperature of $80 \, °C$, at a confinement pressure of 200 bar, a pressure in the pore circuit, containing simulated pore fluid of 100 bar under static conditions, and a pressure in the drilling fluid circuit, which in this phase contains simulated pore fluid, of 100 bar under flow conditions.

Evaluation of the Drilling Fluid

The pore fluid, circulating in the mud circuit, is changed over with test fluid that is formulated with salts or with salts and clay inhibitor additives. When all the pore fluid circulating in the mud circuit has been replaced by the test fluid, the pressure in the mud circuit is raised and is maintained constant for the whole duration of the test. The effectiveness of the test fluid in preventing pressure transmission from the mud to the shale is evaluated by the pressure trend in the pore circuit over a period of time and in particular at its equilibrium value.

3.8 Corrosion Inhibitors

Some solid weighting materials can adversely affect the rheological properties of drilling fluids [205]. Therefore, soluble weighting materials have been used, in particular formates, acetates, and propionates of potassium and sodium. These salts exhibit a high solubility in water, and it is possible to prepare from them aqueous solutions of high density which nevertheless have a low viscosity.

Alkali metal carboxylates, however, may have a serious disadvantage. When the drilling passes through rock strata which comprise naturally occurring inorganic ion exchangers, such as zeolites, free carboxylic acids form from the carboxylates. The free carboxylic acids can lead to considerable corrosion on the drilling tools, in particular, when the operation is carried out at a high temperature.

3.8.1 Amido Amine Salts

Amido amine salts have been found to effect an effective corrosion inhibition that can be utilized in formate solutions, inasmuch as the high pH environment does not hydrolyze the amido group [206]. Such a compound may be prepared from inter alia soya fatty acid and N-ethylethylenediamine. The main components of soya fatty acid are linoleic acid, oleic acid, and palmitic acid. These compounds are shown in Figure 15.

3.8.2 Boron Compounds

It has been found that the addition of boron compounds to alkali metal carboxylate containing drilling fluids greatly decreases their corrosivity to metallic materials [205].

Suitable boric acids are orthoboric acid, metaboric acid, and complex boric acids which can be obtained by the addition of polyhydric alcohols, e.g., mannitol, to orthoboric acid.

Certain organic boron compounds do not have the tendency to a spontaneous decomposition in an aqueous medium. These are esters of boric acid with monohydric or dihydric alcohols, such as trimethyl borate, boric acid monoethanol amine ester, or boric acid triethanol amine ester [205].

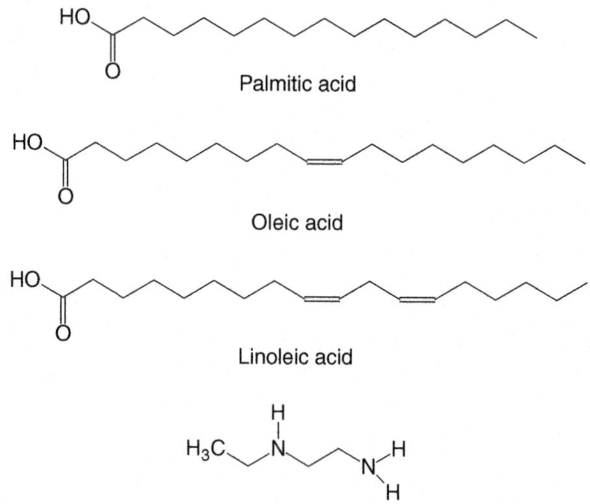

Figure 15 Compounds for amido amine salts.

3.8.3 Phenolic Corrosion Inhibitors

Phenolic compounds are suitable corrosion inhibitors [207]. Such compounds are summarized in Table 27.

Phenolic corrosion inhibitors are shown in Figure 16.

Table 27 Phenolic corrosion inhibitors [207]

Compound	Compound
Phenol	Catechol
o-Cresol	Resorcinol
o-Fluorophenol	Guaiacol
o-Chlorophenol	Hydroquinone
o-Bromophenol	Phloroglucinol
o-Iodophenol	Pyrogallol
o-Nitrophenol	4,4′-Biphenol
o-Allyl phenol	1,3-Dihydroxynaphthalene
Salicylaldehyde	

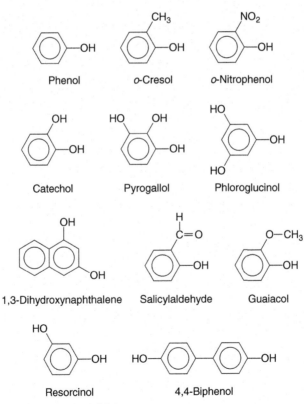

Figure 16 Phenolic corrosion inhibitors.

3.9 Anti-accretion Additives

The protection of the environment from pollutants becomes increasingly important [208]. The toxicity of oil-based fluids can be reduced by the replacement of diesel oil with low-aromatic mineral oils. OBMs may be used but not to be discharged in offshore or inland waters.

The cleanup of OBMs is costly. Therefore, alternatives to OBMs are of importance. As an alternative, water-based drilling fluids containing gypsum, lime, and partially hydrolyzed PAM are used as replacements for OBMs.

However, there are interactions of the aqueous phase of the fluid with the drill bit, such as bit balling, which occurs when the adhesive forces between the bit surface and drill chip become greater than the cohesive forces holding the chip together [208]. In order to decrease bit balling, the adhesive forces between the deformable shale chip and the bit surface must be reduced. A copolymer from poly(propylene glycol) has been developed to reduce the bit balling in water-based drilling fluids.

A polycrystalline-diamond-compact-bit has been developed [209]. This design comprises a hydraulic layout that optimizes bit cleaning and cuttings removal in soft and sticky formations. Significant improvements in performance have been achieved in Cretaceous and Triassic formations drilled with WBMs.

Cuttings accretion are problems encountered when drilling shales, particularly with WBMs [210]. Shale cuttings can adhere to each other, to the bottom hole assembly, and to the cutting surfaces of the bit. In time, a large plastic mass builds up which can block the mud circulation and reduce the rates of penetration.

When the cuttings are exposed to a conventional WBM, they usually imbibe water and can pass rapidly through the different zones, eventually dispersing. Recent advances in drilling fluid technology have developed highly inhibitive muds which appear to reduce the hydration of the shale and maintain the cuttings in the plastic zone thus contributing to an increased accretion and bit balling.

Anti-accretion additives are reducing the accretion and bit balling tendency of shale cuttings [210]. These additives are based on phosphonates. Examples are shown in Table 28.

Other additives that reduce the bit balling are alkylsulfosuccinate salts [211]. The additive is primarily classified as a surfactant. In particular,

Table 28 Anti-accretion additives [210]
Compounds

Hydrolyzed polymaleic acid and 3-phosphonopropionic acid
Succinic acid and propyl phosphonic acid
Dibutyl-butyl phosphonate 2-hydroxyphosphonoacetic acid
Dimethylpropyl phosphonate and phosphorous acid
Diethyl-ethylphosphonate and ethyl methacrylate phosphate
Triethyl phosphonoacetate
Tetramethyl phosphonosuccinate
Phosphonosuccinic acid
2-Hydroxyethyl phosphonic acid

2-Hydroxyethyl phosphonic acid   2-Hydroxyphosphonoacetic acid

Ethylmethacrylate phosphate   Phosphonosuccinic acid

3-Phosphonopropionic acid   Triethyl phosphonoacetate

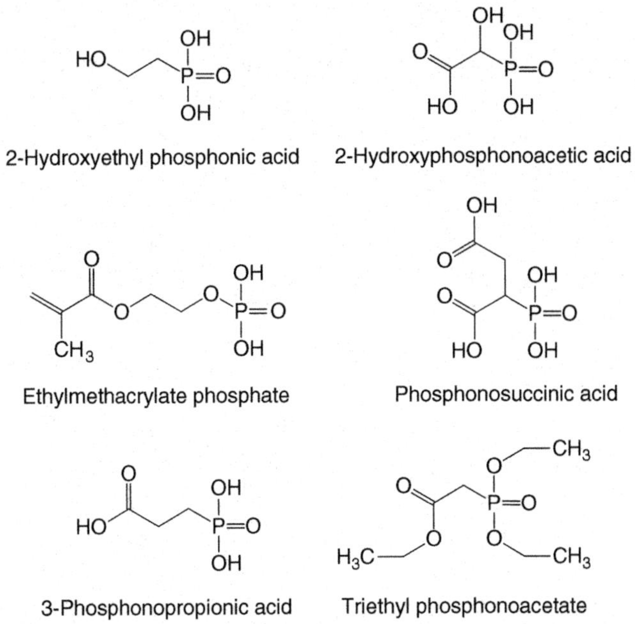

Figure 17 Anti-accretion additives.

sodium diisobutylsulfosuccinate, and sodium dihexylsulfosuccinate have been suggested. These sodium alkylsulfosuccinates are at least to an amount of 25% soluble or dispersible in water and are environmentally nontoxic. Some of these above-mentioned compounds are shown in Figure 17.

4. SPECIAL ISSUES OF WATER-BASED DRILLING FLUIDS

4.1 Formation Damage

The formation damage associated with water-based drilling fluids and emulsified acids is known [212]. There are six scenarios in which drilling fluids can cause a damage of the formation [213]:

1. fluid-to-fluid incompatibilities, e.g., emulsions generated between invading OBM filtrate and formation water;
2. rock-to-fluid incompatibilities, e.g., contact of potentially swelling smectite clay or deflocculatable kaolinite clay by nonequilibrium aqueous fluids, such as fresh water, that have the potential to reduce near-wellbore permeability severely;
3. solids invasion, e.g., the invasion of weighting agents or drilling cuttings into the formation;
4. phase trapping/blocking, e.g., the invasion and entrapment of water-based fluids in the near-wellbore region of gas wells;
5. chemical adsorption/wettability alteration, e.g., changes in the wettability and fluid flow characteristics in the critical near-wellbore area because of emulsifier adsorption;
6. biological activity, e.g., the introduction of bacteria into the formation during drilling and the subsequent generation of slimes, which reduce the permeability.

Moreover, in over-balanced drilling, solids invade the formation [214]. These invading particles, which are suspended in the drilling fluid, tend to plug the pore throats and cause a formation damage. To minimize the formation damage, a properly sized bridging material should be large enough not to invade the formation. Further, it should form an effective filter cake to prevent the invasion of solids and mud–filtrate [215].

A model has been developed that indicates that solid particles with a smaller diameter than the pore throat size will enter the rock and reduce the permeability [216]. The depth of invasion of the solids in drill-in fluids give information on the required flow-initiation pressure.

Manganese Tetroxide

Manganese tetraoxide Mn_3O_4 has been recently used as a weighting material for water-based drilling fluids. Mn_3O_4 has a specific gravity of $4.8\,g\,cm^{-3}$. It is used in muds for drilling deep gas wells. The filter cake formed by this mud contains also Mn_3O_4 [217].

Several articles engaged with the use of manganese tetroxide with other additives in drilling fluid formulations reported negative effects on the reservoir performance. The permeability of reservoirs is reduced when they are contacted with such drilling fluids. Special and expensive stimulation techniques have been proven to be necessary.

Unlike $CaCO_3$, Mn_3O_4 is a strong oxidant [217]. Therefore, the use of HCl is not recommended for the removal of the filter cake. Various organic acids, chelating agents, and enzymes have been tested up to 150 °C.

A research has been presented that indicates that a drilling fluid formulation containing manganese tetroxide with a minimal reduction of the permeability of the reservoir formation with respect to hydrocarbon flow [218]. These formulations are particularly useful for the use in wells that are otherwise difficult to stimulate. A return permeability of 90% or greater is achieved without the need for acidizing treatments.

In order to achieve these performance levels, the formulation must possess certain rheological, density, temperature, and fluid loss properties. Recall that manganese tetroxide has a density of $4.7 \, g \, cm^{-3}$.

A manganese tetroxide-containing wellbore drilling fluid formulation which generates a 90% or greater return permeability without the need for acidizing washes after completion. The formulation is water based. An example of a formulation is given in Table 29.

In comparison to a synthetic mud based on alkalis salts of formic acid, where a return permeability is 66% of the initial volume of oil injected, in a manganese tetroxide based mud a return permeability is 93% [218].

Solid-Free Composition

Formation damage may occur when solids or filtrate derived from a fluid invades the formation during drilling [220]. Such a formation damage can be minimized by treating the well in a near-balanced condition, i.e., the wellbore pressure is close to the formation pressure. However, in

Table 29 Manganese tetroxide formulation [219]

Compound	Amount (lbs bbl^{-1})
Water (fresh) 95.2%	0.952
Bentonite	4.0
Dry Xanthan biopolymer	1.5
Dextride (modified starch with biocide)	6.0
Hydrated lime	0.25
Manganese tetroxide	80.0

high-pressure wells, the well must be often treated in an over- or under-balanced condition. If the well is over-balanced, the treating fluid is designed to temporarily seal the perforations so to prevent the entry of fluids and solids into the formation. In contrast, if the well is under-balanced, the treating fluid is designed to prevent the entry of solids from the formation into the wellbore.

Conventionally, under-balanced drilling technologies use air, foams, and aerated fluids. These air-based technologies solve some of the problems encountered drilling in under-balanced conditions. However, air-based technologies are very costly [220].

Low-density fluids have the capacity to control both over-balanced and under-balanced formations. Water-based low-density fluids are of particular interest, because of their environmental superiority and their relatively low cost. In order to be effective, low density drilling fluids must have adequate rheology and fluid loss control properties. Viscosity and fluid loss may be controlled in low density fluids by adding certain polymers to the fluid. Solid bridging agents also generally are required to prevent fluid loss.

Certain surfactant-polymer blends are effective to generate effective rheology comprising low-shear rate viscosity and effective fluid loss control properties. The water-based fluids preferably are substantially solid free [220].

Surfactant screening for an isomerized olefin internal phase in-water and high molecular weight polyglycol-in-water emulsification showed that the most effective surfactant was sodium tridecyl ether sulfate [220]. This surfactant generated emulsions with initial droplet size of less than 10 μm, which remained stable after hot rolling at 93 °C. However, the droplet size increased considerably to 78 μm, when the emulsions were hot rolled at 120 °C. The increased droplet size can be overcome by using a higher concentration of surfactant or by mixing the surfactant with other additives, such as a polymer that could improve the thermal stability of the drilling fluid [220].

4.2 Uintaite for Borehole Stabilization

Uintaite is a naturally occurring hydrocarbon mineral classified as an asphaltite. It is a natural product whose chemical and physical properties vary and depend strongly on the uintaite source. Uintaite has also been called Gilsonite. Gilsonite is a registered trademark of American Gilsonite Co., Salt Lake City, Utah.

General purpose Gilsonite brand resin has a softening point of about 175 °C, and Gilsonite HM has a softening point of about 190 °C, and Gilsonite Select 300 and Select 325, which have softening points of about 150 °C and 160 °C, respectively. The softening points of these natural uintaites depend primarily on the source vein that is mined when the mineral is produced.

Uintaite is described in Ref. [221]. Typical uintaite used in drilling fluids is mined from an area around Bonanza, Utah, and has a specific gravity of 1.05 with a softening point ranging from 190 to 205 °C, although a lower softening point (165 °C) material is sometimes used. It has a low-acid value, a zero iodine number, and is soluble or partially soluble in aromatic and aliphatic hydrocarbons, respectively.

For many years, uintaite and other asphaltic-type products have been used in water-based drilling fluids as additives assisting in borehole stabilization. These additives can minimize hole collapse in formations containing water-sensitive, sloughing shales. Uintaite and asphalt-type materials have been used for many years to stabilize sloughing shales and to reduce borehole erosion. Other benefits derived from these products include borehole lubrication and reduction in filtration.

Uintaite is not easily water-wet with most surfactants. Thus, stable dispersions of uintaite are often difficult to achieve, particularly in the presence of salts, calcium, solids, and other drilling fluid contaminants and in the presence of diesel oil. The uintaite must be readily dispersible and must remain water-wet; otherwise, it will coalesce and be separated from the drilling fluid, along with cuttings at the shale shaker or in the circulating pits. Surfactants and emulsifiers are often used with uintaite drilling mud additives.

Uintaite is not water-wettable. Loose or poor bonding of the surfactant to the uintaite will lead to its washing off during use, possible agglomeration, and the removal of uintaite from the mud system with the drilling wastes. Thus, the importance of the wettability, rewettability, and storage stability criteria is evident [222].

An especially preferred product comprises about 2 parts Gilsonite HM, about 1 part Gilsonite Select, about 1 part causticized lignite, and about 0.1–0.15 part of a nonionic surfactant [222–225].

Hydrophobic asphaltite can be mixed with a surfactant or dispersant. This mixture is then sheared under a sufficiently high mechanical shear for a sufficient time to convert the hydrophobic asphaltite into hydrophilic asphaltite [226]. Here, a preferred surfactant is ethoxylated glycol. Dispersants can be selected from potassium hydroxide, sodium hydroxide, or lignites.

Sodium Asphalt Sulfonate

Neutralized sulfonated asphalt (i.e., salts of sulfonated asphalt and their blends with materials such as Gilsonite, blown asphalt, lignite, and mixtures of the latter compounds) are commonly used as additives in drilling fluids. These additives, however, cause some foaming in water or water-based fluids. Furthermore, these additives are only partially soluble in the fluids.

Therefore, liquid additives have been developed to overcome some of the problems associated with the use of dry additives. On the other hand, with liquid compositions containing polyglycols, stability problems can arise. Stable compositions can be obtained by special methods of preparation [227]. In particular, first the viscosifier is mixed with water, then the polyglycol, and finally the sulfonated asphalt is added.

4.3 Clay Swelling Inhibitors

Water-based drilling fluids are generally considered to be more environmentally acceptable than oil- or synthetic-based fluids. However, this type of drilling fluid facilitates clay hydration and swelling. Clay swelling, which occurs in exposed sedimentary rock formations, can have an adverse impact on drilling operations and may lead to significantly increased oil well construction costs [228].

For this reason, minimizing clay swelling is an important field of research. In order to reduce the extent of clay swelling effectively, the mechanism of swelling needs to be understood. Based on this knowledge, efficient swelling inhibitors may be developed.

Suitable clay swelling inhibitors must both significantly reduce the hydration of the clay, and must meet the increasingly stringent environmental guidelines.

It takes place in a discrete fashion, namely in the stepwise formation of integer-layer hydrates. The transitions of the distances of the layers are thermodynamically analogous to phase transitions. Electro-osmotic swelling can occur only in clay minerals that contain exchangeable cations in the interlayer region. This type of swelling may yield significantly larger expansion than crystalline swelling.

Sodium-saturated smectites have a strong tendency towards electro-osmotic swelling. In contrast, potassium-saturated smectites do not swell in such a manner. Thus allowing an appropriate ion exchange reaction can be helpful in clay stabilization [228].

The water desorption isotherms of montmorillonite intercalated with exchangeable cations of the alkali metal group showed that for larger cations, less water is adsorbed [229]. In addition, there is a relationship with the tendency to swell and the energy of hydration of the cation [230].

Clay swelling during the drilling of a subterranean well can have a tremendous adverse impact on drilling operations. The overall increase in bulk volume accompanying clay swelling impedes removal of cuttings from beneath the drill bit, increases friction between the drill string and the sides of the borehole, and inhibits formation of the thin filter cake that seals formations. Clay swelling can also create other drilling problems such as loss of circulation or stuck pipe that slows down drilling and increases the drilling costs [231].

In the North Sea and the US Gulf Coast, drillers commonly encounter argillaceous sediments in which the predominant clay mineral is sodium montmorillonite, commonly called gumbo clay. Sodium cations are predominately the exchangeable cations in gumbo clay. Because the sodium cation has a low positive valence (i.e., a +1 valence), it easily disperses into water.

Consequently, gumbo clay is notorious for its swelling. Thus, given the frequency in which gumbo clay is encountered in drilling subterranean wells, the development of a substance and method for reducing clay swelling is of primary importance in the drilling industry [232].

Clays or shales have the ability to absorb water, thus causing the instability of wells either because of the swelling of some mineral species or because the supporting pressure is suppressed by modification of the pore pressure. The response of a shale to a water-based fluid depends on its initial water activity and on the composition of the fluid.

The behavior of shales can be classified into either deformation mechanisms or transport mechanisms [233]. Optimization of mud salinity, density, and filter cake properties is important in achieving optimal shale stability and drilling efficiency with WBM.

Kinetics of Swelling of Clays

Basic studies on the kinetics of swelling have been performed [234]. Pure clays (montmorillonite, illite, and kaolinite) with polymeric inhibitors were investigated, and phenomenologic kinetic laws were established.

Hydrational Stress

Stresses caused by chemical forces, such as hydration stress, can have a considerable influence on the stability of a wellbore [235]. When the total pressure and the chemical potential of water increase, water is absorbed into the clay platelets.

This results either in the platelets moving farther apart (swelling) if they are free to move or in generation of hydrational stress if swelling is constrained [236]. Hydrational stress results in an increase in pore pressure and a subsequent reduction in effective mud support, which leads to a less stable wellbore condition.

Saccharide Derivatives

A drilling fluid additive, which acts as a clay stabilizer, is the reaction product of methyl glucoside and alkylene oxides such as EO, PO, or 1,2-butylene oxide. Such an additive is soluble in water at ambient conditions, but becomes insoluble at elevated temperatures [237]. Because of their insolubility at elevated temperatures, these compounds concentrate at important surfaces such as the drill bit cutting surface, the borehole surface, and the surfaces of the drilled cuttings.

Drilling fluids with a reaction product of sorbitol and alkylene oxide have been found in laboratory tests to exhibit improved shale inhibition properties as compared with known polyol containing WBM, particularly in the absence of potassium ions [238]. This is environmentally advantageous. The alkylene oxide can be EO, PO, or butylene oxide and also mixtures from these alkylene oxides. The reaction products of polyhydroxyalkanes and alkylene oxides can be readily synthesized by ordinary polymerization reactions, e.g., base catalyzed polymerization. The drilling fluid composition may be otherwise a conventional formulation.

It has been argued that the mechanism of shale inhibition may result from an enhanced hydrophobic interaction between adjacent polyol additive molecules adsorbed on clay surfaces of shales due to the increased hydrophobicity of the polyol resulting from the presence of the polyhydroxyalkane. An alternative explanation is that these molecules are effective at disrupting the organization of water molecules near the surfaces of clay minerals. This organization has been proposed as a mechanism for the swelling of clay minerals in aqueous fluids [238].

4.4 Removing Solids

Deposits in a well equipment are typically finely divided inorganic particles, such as solids produced from the formation, which may include hydraulic fracturing proppant, formation sand, clay, and various other precipitates [239]. These particles become coated with hydrocarbonaceous materials and subsequently accumulate additional quantities of heavy hydrocarbonaceous material in the flowlines, settling tanks, and other surfaces on and in the well equipment.

Often, the deposits are slimy, oily substances which strongly adhere to metal and ceramic surfaces, as well as other surfaces of the well equipment, and serve to inhibit fluid flow throughout a drilling or production system.

Common apparatus for mechanically removing solids from drilling mud are shale shakers and vibratory screens, desanders, desilters, and centrifuges. However, mechanical cleaning is often not practicable.

A chemical method of cleaning is to include potassium sulfate as an additive to the well fluid. Still another method and composition for cleaning and inhibiting solid, bitumen tar, and viscous fluid deposition in well a equipment that is used in conjunction with a water-based well fluid has been described. Here, into the well fluid a a miscible terpene is mixed [239]. Examples of suitable terpenes are summarized in Table 30.

Terebene is a mixture of terpenes. D–Limonene, or R–1–methyl–4–(1–methylethenyl) cyclohexene, is a biodegradable solvent occurring in nature as the main component of citrus peel oil. The structure is given in Figure 18.

D–Limonene in water-based drilling fluids improves the penetration rate, and has high lubricity and low toxicity [240].

It has been found that unsaturated terpenes can be converted by hydrogenation into compounds that pass LC-50 toxicity tests. Hydrogenated terpenes and blends of hydrogenated terpenes and unhydrogenated terpenes produce drilling fluids that compare favorably with drilling fluids containing only unhydrogenated terpenes [241]. In Figure 19, the effect of

Table 30 Terpenes for cleaning [239]

Compound	Compound	Compound
D-Limonene	Terebene	
Dipentene		
Pinene	Terpinene	Myrcene
Terpinolene	Phellandrene	Fenchene

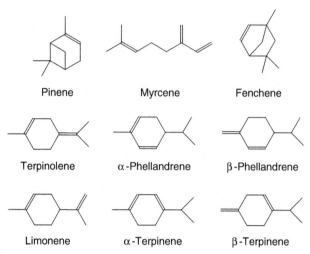

Figure 18 Terpenes.

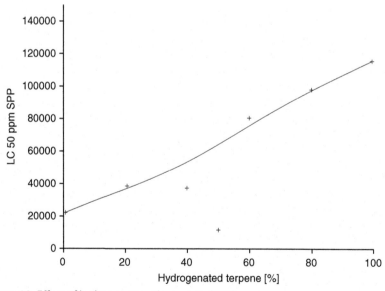

Figure 19 Effect of hydrogenation the toxicity [241].

hydrogenated terpene compounds on the toxicity of terpene drilling fluids is shown.

LC-50 indicates that the suspended particulate phase concentration that will kill 50% of the Mysid shrimp (*Mysidopsis bahia*) used for testing under certain conditions. The vertical axis of the graph is the amount in parts per million of the suspended particulate phase concentration that will kill 50% of the subjects.

4.5 Removing Water-based Drilling Fluids from the Surfaces in a Wellbore

Compositions for removing a water-based drilling fluid from the surfaces in a wellbore and leaving the surfaces water-wet have been described [242].

A water-based drilling fluid removal solution is a mixture of a sulfonated kraft lignin salt and an *N*-methyl-*N*-oleyltaurine salt dissolved in water.

Eventually, the drilling fluid is displaced from the wellbore and the wellbore surfaces are contacted with the removal solution to in order to remove the drilling fluid from the surfaces and make the surfaces water-wet [242].

4.5.1 Shale Reservoirs

The drilling fluid in unconventional shale gas reservoirs has been often selected from a nonaqueous-based fluid [243]. This type can provide advantages, such as shale stabilization, lubricity, and contamination tolerance; however, disadvantages are environmental consequences and associated costs.

For this reason, WBMs for drilling are basically of interest. Since the shale mineralogy and the bottom hole temperature are highly variable critical factors in unconventional gas reservoirs, a single water-based fluid seems to be an unrealistic option.

However, a water-based fluid can be tailored in various aspects. The formulation of a water-based fluid relies on the detailed analysis of the well parameters, such as shale morphology, lithology, planned method of drilling program plans, and environmental factors. In a study, a full laboratory development and testing have been done. In addition, field trials have been done to demonstrate that specially designed water-based fluids can provide a performance comparable to that of nonaqueous-based fluids; however, the environmental and economic benefits are enhanced [243].

Shale Inhibition

Many of the problems arising with the use of water-based fluids in drilling and completion operations are caused by incompatibilities between the fluids and the shales [244]. Such incompatibilities may result in washouts, increased drilling costs, and shale sloughing.

The unstable tendency of water-sensitive shales can be related to water adsorption and hydration of clays. When a WBM comes in contact with shales, water adsorption occurs immediately. This causes clays to hydrate and swell resulting in stress or volume increases. Stress increases can induce brittle or tensile failure of the formations, leading to sloughing, cave-in, and stuck pipe. Volume increases, on the other hand, reduce the mechanical strength of shales and cause swelling of wellbore, disintegration of cuttings in drilling fluid, and balling up of drilling tools. The best way to minimize these drilling problems is to prevent the water adsorption and the clay hydration [245].

Most undesirably is the presence of reactive shales in the region that should become gravel packed [244]. The importance of shale inhibition and the problems associated with shale reactivity during gravel packing are not well known. The selection of a shale inhibitor is often based on the comparison of the results from bottle-roll tests in the laboratory. These tests are admittedly highly functional, but they can give only information on the relative performance of the fluids.

Guidelines for the selection of shale inhibitors for use in gravel packing applications have been worked out. Recommendations are based on the brine type and density, type of shale, temperature, fluid exposure history, and environmental considerations [244].

One potential mechanism by which polymers may stabilize shales is by reducing the rate of water invasion into the shale. The control of water invasion is not the only mechanism involved in shale stabilization [246]. There is also an effect of the polymer additive. Osmotic phenomena are responsible for water transport rates through shales.

Shale Encapsulation A shale encapsulator can be added to a WBM in order to reduce the swelling of the subterranean formation in the presence of water. A shale encapsulator should be at least partially soluble in the aqueous continuous phase in order to be effective.

Inhibiting Reactive Argillaceous Formations Argillaceous formations are very reactive in the presence of water. Such formations can be stabilized by bringing them in contact with a polymer solution with hydrophilic and hydrophobic links [247–249]. The hydrophilic portion consists of poly(oxyethylene), with hydrophobic end groups based on isocyanates. The polymer is capable of inhibiting the swelling or dispersion of the argillaceous rock resulting from its adsorptive and hydrophobic capacities.

Thermal Treatment To increase the permeability of a certain region of the reservoir, the liquid-absorbed water is evaporated by heating the portion to a temperature above the boiling point of water, taking into account the ambient pressure [250, 251]. The liquid water is evaporated by injecting a water-undersaturated gas, such as heated nitrogen, into the reservoir.

Quaternary Ammonium Salts Choline salts are effective anti-swelling drilling fluid additives for under-balanced drilling operations [252]. Choline is addressed as a quaternary ammonium salt containing the *N,N,N*-trimethylethanolammonium cation. Actually, choline is a water-soluble essential nutrient [253]. An example of choline halide counter ion salts is choline chloride.

Preparation II–4: Triethanol amine methyl chloride can be prepared by adding to triethanol amine in aqueous solution methyl chloride in excess and heating for several hours. Upon completion of the reaction, the excess of methyl chloride is evaporated.

Choline formate is prepared from an aqueous solution of choline hydroxide by the reaction with formic acid simply by stirring. ■

Argillaceous formations contain clay particles. If a water-based drilling fluid is used in such formations, then ion exchange, hydration, etc. will take place. These reactions cause swelling, crumbling, or dispersion of the clay particles. Ultimately, washout and even complete collapse of the borehole may occur [254].

Certain additives may prevent these unfavorable reactions. These additives are essentially quaternized polymers. Such polymers have been shown in laboratory testing to vastly reduce shale erosion. Quaternized polymers can synthesized by [254]:

1. quaternization of an AA-based amine derivative with an alkyl halide, and subsequent polymerization, or
2. first polymerization and afterwards quaternization of the polymeric moieties.

Preparation II–5: A quaternized monomer can be prepared by mixing dimethyl amino ethyl methacrylate with hexadecyl bromide. The mixture is heated to 43 °C and stirred for 24 h. Then, the mixture is poured into petroleum ether, whereby the quaternized monomer precipitates [254]. The reaction is shown in Figure 20. ■

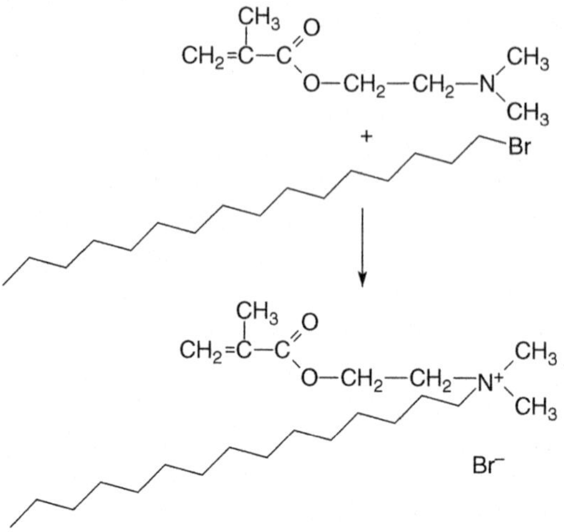

Figure 20 Quaternization reaction of dimethyl amino ethyl methacrylate with hexadecyl bromide.

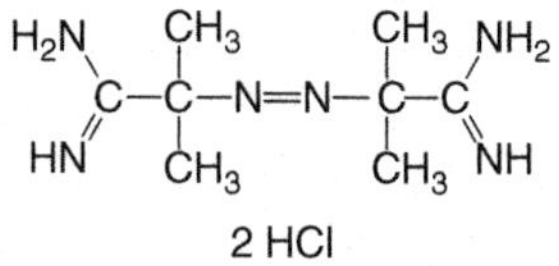

2,2-Azobis (2-amidinopropane) dihydrochloride

Figure 21 Water soluble radical initiator.

A copolymer can be prepared using the quaternized monomer, described above and the dimethyl amino ethyl methacrylate as such. The aqueous solution is neutralized with sulfuric acid and radically polymerized with 2,2′-azobis(2-amidinopropane)dihydrochloride, c.f. Figure 21. This initiator is water soluble. The polymerization is carried out at 43 °C for 18 h [254].

The quaternization of a polymer from dimethyl amino ethyl methacrylate has been described. To an aqueous solution of a homopolymer from dimethyl amino ethyl methacrylate sodium hydrochloride is added to adjust the pH to 8.9. Then again some water is added and hexadecyl bromide as alkylation agent, further benzylcetyldimethyl ammonium bromide as emulsifier. This mixture is then heated, with stirring, to 60 °C for 24 h [254].

4.6 Filter Cake Characterization

The characterization of the filter cake is very important in drilling and completion operations [255]. The homogeneity of the filter cake affects the properties of the filtration process, such as volume of filtrate, and thickness of the filter cake. Also the best method to remove it can be assessed from these data.

Various models have been used to determine the thickness and permeability of a filter cake. Most of these models are based on the assumption of a homogeneous filter cake was.

However, a recent study indicated that the filter cake is not homogeneous, but it consists of two layers with different properties [255].

Computer tomography scan was used to measure the thickness and the porosity of a filter cake. In additions, scanning electron microscopy was used to get information of the morphology of the filter cake.

In a study, a high-pressure high-temperature filter cake was used for testing. The results indicated that the filter cake was heterogeneous and contained two layers with different properties under static and dynamic conditions. Under static conditions, the layer close to the rock surface is 0.06 in. thickness with a porosity of 10-20 vol.%. Under dynamic conditions, this layer is 0.04 in. thickness with a porosity of 15 vol.%. In contrast, the layer close to the drilling fluid is 0.01 in. thickness under static conditions, and 0.01 in. thick under dynamic conditions.

Scanning electron microscopy showed that the two layers contain both large and small particles, but an extremely poor sorting in the layer is found, that is close to the drilling fluid. This results in a nearby zero porosity in this layer.

It was stated by the authors that previous models underestimated the thickness of the filter cake by almost 50% [255].

4.7 Differential Sticking

A drilling fluid typically comprises a liquid carrier, usually water or brine, but sometimes also oil, a viscosifier which is often bentonite clay, a weighting agent such as barite, and a number of additives for modifying the properties of the fluid [256].

In the course of drilling, a certain amount of liquid phase can sometimes permeate into the drilled rock causing the solids to be filtered at the borehole wall so as to form a filter cake or mud cake. The mud cake prevents further fluid loss to some degree. However, if the mud cake continues to grow, there is a danger of differential sticking of the drill pipe [256].

Differential sticking is particularly likely to occur when drilling high angle holes from offshore platforms [257]. It happens when a portion of the drill string lies against a side of a deviated hole where a filter cake has built up adjacent a permeable formation.

While the pipe is being rotated, it is lubricated by a film of mud, and the hydrostatic pressure exerted by the drilling fluid is equal on all sides of the pipe. However, when rotation of the pipe is stopped, the portion of the pipe in contact with the filter cake is isolated from the mud column, and the differential of the hydrostatic pressure of the mud and the formation pressure exerted on opposing sides of the pipe presses the pipe into the filter cake and causes drag when an attempt is made to pull the pipe.

The force required to pull free is the product of the coefficient of friction u and the force pushing the drill string against the formation. The pushing force is the product of the contact area A between the drill pipe and the cake, and the excessive hydrostatic pressure p at the contact area, which is equal to the difference between the pressure of the mud in the hole and the formation pressure. Thus, the force F to pull free is [257]:

$$F = uAp. \tag{1}$$

The coefficient of friction u between the steel drill pipe and the mud cake depends on the composition of the mud. The area of wall contact depends not only on the diameters of the pipe and the borehole, but also on filter cake thickness. The force to pull free may exceed the power of surface equipment, and unless the stuck pipe can be released, the hole, the drill string, and equipment on the drill string below where the pipe is stuck may be lost. This can cause expensive remedial measures and usually side track (directional) drilling above the stuck pipe if that much of the hole is not to be abandoned [257].

Differential sticking occurs when the mud cake of a water-based fluid grows so as to contact the drill pipe and will typically occur when the pipe is not rotating or otherwise moving. This happens to a much less degree with oil-based fluids. The pipe sticks when the torque or overpull which is available is insufficient to free the pipe from the mud cake [256].

In the past, oil has been used to loosen stuck drill pipe and to lubricate the borehole before running casing [257]. An oil slug weighted to the same density of the WBM, i.e., a *pill*, is placed in and circulated through the mud system in volume adequate to extend from the borehole bottom to a level at least as high as the uppermost point of differential sticking. The oil fluid invades the mud cake, reduces the adhesive forces, and lubricates the pipe to reduce friction and facilitate a release of the pipe.

Later oil emulsions and OBMs have been favored for preventing differentially stuck drill pipe. Oil emulsion and OBMs have much lower coefficients of friction than WBMs, and they also lay down very thin filter cakes, which reduces the contact area. However, environmental concerns enforced by world-wide governmental regulations increasingly have limited the use of oil emulsion or oil-based formulations, or oil spotting fluids in drilling operations. The mud might escape into environmentally sensitive areas, such as offshore waters. Environmental regulations for offshore drilling fluids require that [257]:

- no sheen be left upon the receiving waters by the drilling fluid;
- the drilling fluid meet stringent toxicity limits as measured by bioassays of *M. bahia* shrimp.

Beyond marine toxicity or other pollution concerns offshore, there is also difficulty and considerable expense in barging and holding large quantities of oil fluids for offshore use. Further, all oil muds, and cuttings contaminated with oil muds, must be contained and transported back to shore for disposal. Moreover, land-based drilling has the expense of disposing of cuttings and used OBMs in an environmentally acceptable manner [257].

WBMs have been described, which contain poly(oxyalkylene glycol) additives to counter the problem of differential drill pipe sticking [258].

Poly(ethylene oxide), EO-PO copolymers, or poly(vinylmethylether) in brine can be used in spotting pills to dehydrate and crack the filter cake to release a stuck pipe [259].

Various additives have been proposed to assist in freeing stuck drill pipe, the most common of which is diesel oil which is added directly to the drilling mud as a spotting fluid. However this is not always successful. An oil-in-water microemulsion has been proposed for avoiding the sticking [256]. A water-based drilling fluid incorporating the micellar phase additive might have a composition shown in Table 31.

Other examples of (micro)emulsion additives have been created by mixing various surfactants with a mineral oil and the mixtures were added to the standard mud and stickance ratio and fluid loss was measured. The surfactants and surfactant/oil ratios are given in Table 32.

It can be seen from Table 32 that anionic surfactants are superior to nonionic or cationic surfactants in providing the required property without affecting the mud rheology.

Table 31 Micellar phase additive [256]

Component	Amount
Anionic surfactant	5% v/v
Co-surfactant	Optional
Inorganic salt	$4\text{-}6\,g\,l^{-1}$
Bentonite clay	$40\text{-}80\,g\,l^{-1}$
Fluid loss control polymer (CMC)	$3\,g\,l^{-1}$
Antifoaming agent	$<1\,g\,l^{-1}$
Weighting agent	As required

Table 32 Properties of emulsion additives [256]

Trade name	Composition	Surfactant/ oil ratio	Stickance ratio	Fluid loss (ml)
	Standard WBM	–	1.0	3.2
	Standard WBM + oil	–	0.9	2.2
Standard WBM + Anionic surfactant				
Petromix 9	Petroleum sulfonate blend	3:2	0.2	0.4
Petronate L	Petroleum sulfonate	3:2	0.31	1.3
Petronate HH	Petroleum sulfonate	3:2	0.26	1.4
Roval IDBS	IDBS	3:2	0.63	1.4
Sulframin 1250	NaDBS	3:1	0.71	1.8
	KDBS	3:2	0.55	1.9
Sulframin AOS	Sodium dodecyl sulfonate	2:3	0.73	2.0
ROVAL 70PG	Dioctyl sulfosuccinate	3:2	0.57	1.5
EMPHOS 20	Organic phosphate ester	2:3	1.81	13.5
PS21A TWEEN 20	Poly(oxyethylene) sorbitan monolaurate	3:2	1.46	6.2
Standard WBM + Nonionic surfactant				
Synperonic 91/4	Alkyl phenol ethoxylate	3:2	0.29	7.0
Rewo RO 40	Ethoxylated castor oil	3:2	2.9	9.2
Standard WBM + Cationic surfactant				
Ethoxamine SF11	Ethoxylated fatty amine	3:2	0.71	3.3
Sochamine 35	Imidazoline	3:2	0.56	8.2

Castor oil is a source of ricinoleic acid that is an unsaturated 18-carbon fatty acid. Sulfonates appear to be the best anionic surfactants. Some surfactants are shown in Figure 22.

Tradenames appearing in the references are shown in Table 33.

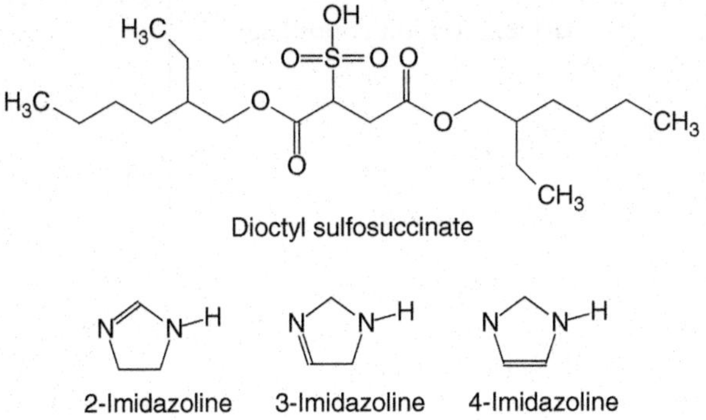

Dioctyl sulfosuccinate

2-Imidazoline   3-Imidazoline   4-Imidazoline

Figure 22 Surfactants.

Table 33 Tradenames in references

Tradename Description	Supplier
Alfonic® 1412–60	Vista Chemical Co.
Ethoxylated linear alcohol [155, 158]	
ALL-TEMP®	Baker Hughes, Inc.
Acrylate tetrapolymer [75]	
AquaPAC®	Aqualon Corp.
Polyanionic cellulose [84]	
Arco Prime™	Lyondell Petrochemical Comp.
Paraffin oil [165]	
Avanel® S150	BASF AG
Anionic surfactant [119]	
Baracarb®	Halliburton Energy Services, Inc.
Ground marble [113]	
Barazan®	Halliburton Energy Services, Inc.
Poly(saccharide) [8, 191]	
Barodense®	Halliburton Energy Services, Inc.
Ground hematite [113]	

Continued

Table 33 Tradenames in references—cont'd
Tradename

Description	Supplier
Bentone® 128 Organically modified smectite, (viscosifier) [82]	Elementis Specialties
BIO-LOSE™ Complexed polysaccharide, filtration control agent [108, 182]	Baker Hughes, Inc.
BIO-PAQ™ Water-soluble polymer [108]	Baker Hughes, Inc.
Biovis® Scleroglucan viscosifier [26, 162, 196]	Messina Chemicals, Inc.
Bore-Drill™ Anionic polymer [75]	Borden Chemicals
Brij® (Series) Ethoxylated fatty alcohols [162]	ICI Surfactants
Brij® 76 Poly(oxyethylene) (10) stearyl ether [162]	ICI Surfactants
Carbo-Gel® Amine modified, gel-forming organophilic clay [165]	Baker Hughes, Inc.
Carbolite™ Sized ceramic proppant [252]	Carbo Corp.
Carbowax® (Series) Poly(ethyleneoxide glycol) (PEG) [158]	Union Carbide Corp.
Cellex® Polyanionic cellulose [191]	CP Kelco
Celpol® (Series) Polyanionic cellulose [84]	Noviant, Nijmegen
CFR™ 3 Cement friction reducer dispersant [113]	Halliburton Energy Services, Inc.
Chek-Loss® PLUS Ultra-fine lignin [75, 108]	Baker Hughes, Inc.
Chemtrol® X Blend of ground lignitic earth and synthetic maleic anhydride copolymers [75]	Baker Hughes, Inc.
Clay Sync™ Shale stabilizer [17, 191]	Baroid Fluid Services
ClaySeal® Shale stabilizer [17, 191]	Baroid Fluid Services
Claytone-II™ Alkyl quaternary ammonium bentonite clay [119]	Southern Products, Inc.

Table 33 Tradenames in references—cont'd
Tradename
Description	Supplier
Claytone® Organophilic bentonite [113]	Claytone
Dacron® Poly(ethylene terephtthalate) [252]	DuPont
DFE-129™ Acrylamide/AMPS copolymer [75]	Baker Hughes, Inc.
DFE-243 Partially hydrolyzed polyacrylamide /trimethylaminoethyl acrylate [108]	Baker Hughes, Inc.
Drill Gel ® Bentonite [159]	CETCO Tech.
DrillAhead® Software [113]	Halliburton Energy Services, Inc.
Driltreat™ Wetting agent [113]	Halliburton Energy Services, Inc.
Driscal® D Water-soluble polymer [75]	Drilling Specialties Comp.
Dualflo® Modified starch [82, 162]	M-I Swaco
Duovis® Xanthan gum [196]	M-I Swaco - Schlumberger
Ecodrill® 317 Potassium silicate [159]	National Silicates, Canada
Ecorr™ RNM 45 Ground rubber [82]	Rubber Resources B.V.
Ecoteric™ Fatty acid ethoxylates [199]	Huntsman
Empol™ (Series) Oligomeric oleic acid [158]	Henkel
Escaid® 110 Petroleum distillate [158]	Exxon Mobil
EZ MUL® NT Emulsifier [113]	Halliburton Energy Services, Inc.
Filter-Chek® Modified Cellulose [8, 17, 191]	Halliburton Energy Services, Inc.
Filterchek™ Modified natural polymer control agent [191]	LS Baltica

Continued

Table 33 Tradenames in references—cont'd
Tradename

Description	Supplier
FlexPlug® OBM	Halliburton Energy Services, Inc.
Reactive, nonparticulate lost-circulation materia [113]	
FlexPlug® W	Halliburton Energy Services, Inc.
Reactive, nonparticulate lost-circulation materia [113]	
Flo-Chek®	Halliburton Energy Services, Inc.
Lost circulation additive [113]	
Flotrol™	M-I Swaco - Schlumberger
Starch derivative [196]	
Flovis™	M-I Swaco
xanthan viscosifier [82]	
Flowzan®	Phillips Petroleum Comp
Xanthan gum [63, 182]	
Glydril®	M-I Swaco - Schlumberger
Poly(glycol) (cloud point additive) [196]	
Grabber®	Baroid Fluid Services
Flocculant [8, 17, 191]	
Hydro-Guard®	Halliburton Energy Services, Inc.
Inhibitive water-based fluid [17]	
Hymod Prima™	Imerys Minerals
Ball clay [82]	
Hyperdrill™ CP-904L	Hychem, Inc.
Acrylamide copolymer [108, 109]	
Hystrene® (Series)	PMC Biogenix, Inc.
Dimer acids [158]	
Imbiber beads®	Imtech Imbibitive Technologies Corp.
Crosslinked alkyl styrene beads [82]	
Inipol™ AB40	CECA
Scale inhibitor [82]	
Invermul®	Halliburton Energy Services, Inc.
Blends of oxidized tall oil and polyaminated fatty acids [113]	

Table 33 Tradenames in references—cont'd

Tradename Description	Supplier
ISO-TEQ isomerized olefins [135]	Baker Hughes, Inc.
Jeffamine® D-230 Poly(oxypropylene) diamine [232]	Huntsman
Jeffamine® EDR-148 Triethyleneglycol diamine [232]	Huntsman
Jeffamine® HK-511 Poly(oxyalkylene) amine [232]	Huntsman
Kasolv® 16 Potassium silicate [161]	PQ Corp.

KEM-SEAL® PLUS
  NaAMPS/N,N–dimethylacrylamide
  copolymer [75]
	Baker Hughes, Inc.

Kemseal® Fluid loss additive [75]	Baker Hughes, Inc.
Klerzyme™ Pectinase [33]	Centerchem
Kuralon™ Poly(vinyl alcohol) fiber [82]	Kuraray, Osaka
Latex™ 2000 Styrene/butadiene copolymer [119]	Halliburton Energy Services
Ligco® Lignite [75]	Baker Hughes, Inc.
Ligcon® Causticized lignite [75]	Milchem Inc.
Magma™ Extrusion spun mineral fiber [82]	Lost Circulation Specialists, Inc.
MAX-TROL® Sulfonated resin [75]	Baker Hughes, Inc.
Microtox® Standardised toxicity test system [197]	AZUR Environmental, MW Monitoring IP, Beckman Instruments, Inc.
Mikhart™ (Series) Fine calcium carbonate particles [82]	M-I Swaco
Mil-Bar® Barite weighting agent [75, 165, 182]	Baker Hughes, Inc.
Mil-Carb® Ground marble [75, 108, 109, 182]	Baker Hughes, Inc.

Continued

Table 33 Tradenames in references—cont'd
Tradename

Description	Supplier
Mil-Gel-NT®	Baker Hughes, Inc.
Bentonite quartz mixture [75]	
Mil-Gel™	Baker Hughes, Inc.
Ground montmorillonite [75, 182]	
Mil-Pac LV	Baker Hughes, Inc.
Low viscosity polyamine cellulose [182]	
Mil-Temp®	Baker Hughes, Inc.
Maleic anhydride copolymer [75]	
N-Dril™ HT Plus	Baroid Fluid Services
Filtration control agent [17]	
Neodol® (Series)	Shell
Alkyl alkoxylated surfactants [155, 158]	
New Drill®	Baker Hughes, Inc.
Partially hydrolyzed polyacrylamide [182]	
NEW-DRILL® PLUS	Baker Hughes, Inc.
Partially hydrolyzed poly(acrylamide) [108, 182]	
Norsorex™ (Series)	Atofina
polynorbornene [82]	
PAC™ -L	Baroid Fluid Services
Filtration control agent [17]	
Paramul™	M-I SWACO - Schlumberger
Primary emulsifier [82]	
Parawet™	M-I SWACO - Schlumberger
Emulsifier [82]	
Pectinol®	Rohm & Haas Comp.
Pectinase [33]	
Petrofree® LV	Halliburton Energy Services, Inc.
Ester based invert emulsion [8]	
Petrofree® SF	Halliburton Energy Services, Inc.
Olefin based invert emulsion [8, 191]	
Pluracol® V-10	BASF AG
Polyoxyalkylene polyol [158]	
Poly-plus® HD	M-I SWACO - Schlumberger
Acrylic copolymer (shale encapsulation) [196]	

Table 33 Tradenames in references—cont'd

Tradename Description	Supplier
Polydrill® Anionic polymer [75]	Degussa AG
Polypac® UL Polyanionic cellulose [196]	M-I SWACO - Schlumberger
Polyplus® RD Acrylic copolymer (shale stabilizator) [196]	M-I SWACO - Schlumberger
Protecto-Magic™ Ground asphalt [75]	Baker Hughes, Inc.
Pyro-Trol® Acrylamide/AMPS copolymer [75]	Baker Hughes, Inc.
Rev Dust Artificial drill solids [75]	Milwhite, Inc.
Shale Guard™ NCL100 Shale anti-swelling agent [252]	Weatherford Int.
Soltex® Sulfonated asphalt [75]	Chevron Phillips Chemical Comp.
Soyafibe™ soybean polysaccharide [34]	Fuji Oil Comp.
Staflo® PAC [84]	Akzo Nobel
Sulfa-Trol® Sulfonated asphalt [75]	Baker Hughes, Inc.
Sunyl® high High oleic sunflower oil [155]	SVO Enterprises
Superfloc™ Acrylamide copolymer [108, 109]	Cytec Industries, Inc.
Surfonamine® Amines as Pigment Modifiers [199]	Huntsman
Surfonic® Emulsifier, ethoxylated C12 alcohol [158]	Huntsman Performance Products
Tergitol® (Series) Ethoxylated C11-15-secondary alcohols, surfactant [155, 158]	Union Carbide Corp.
Triton® N (Series) Alkyl-aryl alkoxylated surfactant [155, 158]	Union Carbide Corp.
Triton® X (Series) Poly(alkylene oxide), nonionic surfactants [155, 158]	Union Carbide Corp. (Rohm & Haas)

Continued

Table 33 Tradenames in references—cont'd

Tradename Description	Supplier
Twaron® Aramid [82]	Teijin Twaron B.V.
Tychem® 68710 Carboxylated styrene/butadiene copolymer [113]	Reichhold
Tylac® CPS 812 Carboxylated styrene/butadiene copolymer [113]	Reichhold
Unamide® Polyoxyalkylated fatty amides [158]	Lonza, Inc.
Versatec™ Oil-based mud [82]	M-I Swaco
Versatrol™ HT Gilsonite [82]	M-I SWACO - Schlumberger
Westvaco® Diacid Diels-Alder acylating agents [158]	Westvaco Corp.
XAN-PLEX™ D Polysaccharide viscosifying polymer [108]	Baker Hughes, Inc.
Xanvis™ Polysaccharide viscosifying polymer [108, 109]	Baker Hughes, Inc.
Ziboxan® Xanthan gum [159]	Shandong Deoson Corp.

REFERENCES

[1] Nzeadibe KI, Perez GP. Method and biodegradable water based thinner composition for drilling subterranean boreholes with aqueous based drilling fluid. US Patent 8453735, assigned to Halliburton Energy Services, Inc. (Houston, TX); 2013. URL: http://www.freepatentsonline.com/8453735.html.

[2] Lyons WC, Plisga GJ, editors. Standard handbook of petroleum and natural gas engineering. 2nd ed. Burlington, MA: Gulf Publishing Co.; 2011 [an imprint of Elsevier]. URL: http://www.sciencedirect.com/science/book/9780750677851.

[3] Van Der Horst PM. Use of CMC in drilling fluids. US Patent 7939469, assigned to Dow Global Technologies LLC; 2011. URL: http://www.freepatentsonline.com/7939469.html.

[4] Gray GR, Darley HCH. Composition and properties of oil well drilling fluids. 4th ed. Houston, TX: Gulf Publishing Co.; 1981. ISBN 0872011291 9780872011298.

[5] Müller H, Herold CP, von Tapavicza S, Fues JF. Use of selected ester oils in water based drilling fluids of the o/w emulsion type and corresponding drilling fluids with improved ecological acceptability. US Patent 5318956, assigned to Henkel Kommanditgesellschaft auf Aktien (DE); 1994. URL: http://www.freepatentsonline.com/5318956.html.

[6] Wu AM, Brockhoff J. Emulsified polymer drilling fluid and methods of preparation. US Patent 8293686, assigned to Marquis Alliance Energy Group Inc. (Calgary, CA); 2012. URL: http://www.freepatentsonline.com/8293686.html.

[7]   Griffin WC. Classification of surface-active agents by HLB. J Soc Cosmetic Chemists 1946;1:311-26.

[8]   West GC, Grebe EL, Carbajal D. Inhibitive water based drilling fluid system and method for drilling sands and other water-sensitive formations. US Patent 7439210, assigned to Halliburton Energy Services, Inc. (Duncan, OK); 2008. URL: http://www.freepatentsonline.com/7439210.html.

[9]   Argillier JF, Audibert-Hayet A, Zeilinger S. Water based foaming composition-method for making same. US Patent 6172010, assigned to Institut Francais du Petrole (Rueil, FR); 2001. URL: http://www.freepatentsonline.com/6172010.html.

[10] Amanullah M, Bubshait AS, Fuwaires OA. Volcanic ash-based drilling mud to overcome drilling challenges. US Patent 8563479, assigned to Saudi Arabian Oil Company (SA); 2013. URL: http://www.freepatentsonline.com/8563479.html.

[11] Lee LJ, Patel AD, Stamatakis E. Glycol based drilling fluid. US Patent 6291405, assigned to M-I LLC (Houston, TX); 2001. URL: http://www.freepatentsonline.com/6291405.html.

[12] Mullen GA, Gabrysch A. Synergistic mineral blends for control of filtration and rheology in silicate drilling fluids. US Patent 6248698, assigned to Baker Hughes Incorporated (Houston, TX); 2001. URL: http://www.freepatentsonline.com/6248698.html.

[13] Charles UJ. Potassium silicate drilling fluid. WO Patent 1997005212 assigned to Urquhart John Charles; 1997. URL: https://www.google.at/patents/WO1997005212A1?cl=en.

[14] Kotelnikov VS, Demochko SN, Fil VG, Marchuk IS. Drilling mud composition – contains carboxymethyl cellulose, acrylic polymer, ferrochrome lignosulphonate, cement and water. SU Patent 1829381, assigned to Ukr. Natural Gas Res. Inst.; 1996.

[15] Gajdarov MM, Tankibaev MA. Nonclayey drilling solution – contains organic stabiliser, caustic soda, water and mineral additive in form of zinc oxide, to improve its thermal stability. RU Patent 2051946, assigned to Aktyubinsk Oil Gas Inst.; 1996.

[16] Muller GT, Patel BB. Compositions comprising an acrylamide-containing polymer and use. EP Patent 0728826 assigned to Phillips Petroleum Company; 2001. URL: https://www.google.at/patents/EP0728826B1?cl=en.

[17] Maresh JL. Wellbore treatment fluids having improved thermal stability. US Patent 7541316, assigned to Halliburton Energy Services, Inc. (Duncan, OK); 2009. URL: http://www.freepatentsonline.com/7541316.html.

[18] Warren B, van der Horst PM, van't Zelfde TA. Quaternary nitrogen containing amphoteric water soluble polymers and their use in drilling fluids. US Patent 6281172, assigned to Akzo Nobel NV (NL); 2001. URL: http://www.freepatentsonline.com/6281172.html.

[19] Berry VL, Cook JL, Gelderbloom SJ, Kosal DM, Liles DT, Olsen Jr CW, et al. Silicone resin for drilling fluid loss control. US Patent 7452849, assigned to Dow Corning Corporation (Midland, MI); 2008. URL: http://www.freepatentsonline.com/7452849.html.

[20] King KL, McMechan DE, Chatterji J. Water based polymers for use as friction reducers in aqueous treatment fluids. US Patent 7271134, assigned to Halliburton Energy Services, Inc. (Duncan, OK); 2007. URL: http://www.freepatentsonline.com/7271134.html.

[21] King KL, McMechan DE, Chatterji J. Methods, aqueous well treating fluids and friction reducers therefor. US Patent 6784141, assigned to Halliburton Energy Services, Inc. (Duncan, OK); 2004. URL: http://www.freepatentsonline.com/6784141.html.

[22] Martyanova SV, Chezlov AA, Nigmatullina AG, Piskareva LA, Shamsutdinov RD. Production of lignosulphonate reagent for drilling muds – by initial heating with sulphuric acid, condensation with formaldehyde, and neutralisation of mixture with sodium hydroxide. RU Patent 2098447, assigned to Azimut Res. Prod. Assoc.; 1997.

[23] Ibragimov FB, Kolesov AI, Konovalov EA, Rud NT, Gavrilov BM, Mojsa JN, et al. Preparation of lignosulphonate reagent – for drilling solutions, involves additional introduction of water-soluble salt of iron, and anti-foaming agent. RU Patent 2106383; 1998.

[24] Patel BB. Tin/cerium compounds for lignosulfonate processing. EP Patent 0600343 assigned to Phillips Petroleum Company; 1998. URL: https://www.google.at/patents/EP0600343B1?cl=en.

[25] Patel BB, Dixon GG. Drilling mud additive comprising ferrous sulfate and poly(n-vinyl-2-pyrrolidone/sodium 2-acrylamido-2-methylpropane sulfonate). US Patent 5204320, assigned to Phillips Petroleum Company (Bartlesville, OK); 1993. URL: http://www.freepatentsonline.com/5204320.html.

[26] Cobianco S, Bartosek M, Guarneri A. Non-damaging drilling fluids. US Patent 6495493, assigned to ENI S.p.A. (Rome, IT) Enitechnologie S.p.A. (San Donato Milanese, IT); 2002. URL: http://www.freepatentsonline.com/6495493.html.

[27] Vaussard A, Ladret A, Donche A. Scleroglucan based drilling mud. US Patent 5612294, assigned to Elf Aquitaine (Courbevoie, FR); 1997. URL: http://www.freepatentsonline.com/5612294.html.

[28] Keilhofer G, Matzinger M, Plank J, Reichenbach-Klinke R, Spindler C. Water-soluble and biodegradable copolymers on a polyamide basis and use thereof. WO Patent 2008019987 assigned to Basf Construction Polymers Gmb, Gregor Keilhofer, Martin Matzinger, Johann Plank, Roland Reichenbach-Klinke, Christian Spindler; 2008. URL: https://www.google.at/patents/WO2008019987A1?cl=en.

[29] Matzinger M, Reichenbach-klinke R, Keilhofer G, Plank J, Spindler C. Water-soluble and biodegradable copolymers on a polyamide basis and use thereof. US Patent Application 20100240802, assigned to BASF; 2010. URL: http://www.freepatentsonline.com/20100240802.html.

[30] Elward-Berry J. Rheologically stable water based high temperature drilling fluids. US Patent 5244877, assigned to Exxon Production Research Company (Houston, TX); 1993. URL: http://www.freepatentsonline.com/5244877.html.

[31] API Standard RP 13B-1. Recommended practice for field testing water based drilling fluids. API Standard API RP 13B-1; American Petroleum Institute; Washington, DC; 2009.

[32] Elward-Berry J. Rheologically-stable water based high temperature drilling fluid. US Patent 5179076, assigned to Exxon Production Research Company (Houston, TX); 1993. URL: http://www.freepatentsonline.com/5179076.html.

[33] Weibel MK. Parenchymal cell cellulose and related materials. US Patent 4831127, assigned to SBP, Inc. (Philadelphia, PA); 1989. URL: http://www.freepatentsonline.com/4831127.html.

[34] Lundberg B, Huppert A. Dairy product compositions using highly refined cellulosic fiber ingredients. US Patent 8399040, assigned to Fiberstar Bio-Ingredient Technologies, Inc. (River Falls, WI); 2013. URL: http://www.freepatentsonline.com/8399040.html.

[35] Hale AH. Water base drilling fluid. GB Patent 2216573, assigned to Shell Internat. Res. Mij BV; 1989.

[36] Hale AH, Blytas GC, Dewan AKR. Water base drilling fluid. GB Patent 2216574, assigned to Shell Internat. Res. Mij BV; 1989.

[37] Grainger N, Herzhaft B, White M, Audibert Hayet A. Well drilling method and drilling fluid. US Patent 7055628, assigned to Institut Francais du Petrole (Rueil Malmaison Cedex, FR) Imperial Chemical Industries Plc. (London, GB); 2006. URL: http://www.freepatentsonline.com/7055628.html.

[38] Mody FK, Fisk Jr JV. Water based drilling fluid for use in shale formations. US Patent 5925598, assigned to Bairod Technology, Inc. (Houston, TX); 1999. URL: http://www.freepatentsonline.com/5925598.html.

[39]   Schlemmer RF. Membrane forming in-situ polymerization for water based drilling fluids. US Patent 7279445, assigned to M-I L.L.C. (Houston, TX); 2007. URL: http://www.freepatentsonline.com/7279445.html.

[40]   Schlemmer R. Membrane forming in-situ polymerization for water based drilling fluid. US Patent 7063176, assigned to M-I L.L.C. (Houston, TX); 2006. URL: http://www.freepatentsonline.com/7063176.html.

[41]   Maillard LC. Reaction generale des acides amines sur les sucres: Ses consequences biologiques. Compt Rend Acad Sci 1912;154:66-8.

[42]   Mody FK, Pober KW, Tan CP, Drummond CJ, Georgaklis G, Wells D. Compounds and method for generating a highly efficient membrane in water based drilling fluids. US Patent 6997270, assigned to Halliburton Energy Services, Inc. (Duncan, OK) Commonwealth Scientific and Industrial Research Organisation; 2006. URL: http://www.freepatentsonline.com/6997270.html.

[43]   Zhang J, Al-Bazali T, Chenevert M, Sharma M. Factors controlling the membrane efficiency of shales when interacting with water based and oil-based muds. SPE Drilling & Completion 2008;23(2). doi:10.2118/100735-PA.

[44]   Hale AH, Loftin RE. Glycoside-in-oil drilling fluid system. US Patent 5494120, assigned to Shell Oil Company (Houston, TX); 1996. URL: http://www.freepatentsonline.com/5494120.html.

[45]   Headley JA, Walker TO, Jenkins RW. Environmentally safe water based drilling fluid to replace oil-based muds for shale stabilization. In: Proceedings Volume. SPE/IADC Drilling Conf. (Amsterdam, The Netherlands, 2/28/95-3/2/95); 1995, p. 605-12.

[46]   Müller H, Stoll G, Herold CP, Von Tapavicza S. Use of selected ethers of monofunctional alcohols in drilling fluids. EP Patent 0391251 assigned to Henkel Kommanditgesellschaft auf Aktien; 1992. URL: https://www.google.at/patents/EP0391251B1?cl=en.

[47]   Godwin AD, Mathys GMK. Ester-free ethers. WO Patent 1993004028 assigned to Exxon Chemical Patents Inc; 1993. URL: https://www.google.at/patents/WO1993004028A1?cl=en.

[48]   Godwin AD, Sollie T. Load bearing fluid. EP Patent 0532128 assigned to Exxon Chemical Patents Inc.; 1993. URL: https://www.google.at/patents/EP0532128A1?cl=en.

[49]   Müller H, Herold CP, Fues JF. Use of surface-active alpha-sulfo-fatty acid di-salts in water and oil based drilling fluids and other drill-hole treatment agents. US Patent 5508258, assigned to Henkel Kommanditgesellschaft auf Aktien (Düsseldorf, DE); 1996. URL: http://www.freepatentsonline.com/5508258.html.

[50]   Müller H, Herold CP, von Tapavicza S. Oleophilic alcohols as components of invert-emulsion drilling fluids. EP Patent 0391252 assigned to Henkel Kommanditgesellschaft auf Aktien; 1993. URL: https://www.google.at/patents/EP0391252B1?cl=en.

[51]   Müller H, Herold CP, von Tapavicza S. Oleophilic alcohols as a constituent of invert drilling fluids. US Patent 5348938, assigned to Henkel Kommanditgesellschaft auf Aktien (DE); 1994. URL: http://www.freepatentsonline.com/5348938.html.

[52]   Müller H, Stoll G, Herold CP, von Tapavicza S. Drilling fluids. ZA Patent 9002665, assigned to Henkel KG auf Aktien; 1990.

[53]   Müller H, Herold CP, von Tapavicza S. Oleophilic basic amine derivatives as additives for invert emulsion drilling fluids. EP Patent 0382070 assigned to Henkel Kommanditgesellschaft auf Aktien; 1993. URL: https://www.google.at/patents/EP0382070B1?cl=en.

[54]   Lafuma F, Monfreux N, Perrin P, Sawdon C. Invertible emulsions stabilised by amphiphilic polymers and application to bore fluids. WO Patent 2000031154 assigned to Schlumberger Ca Ltd, Schlumberger Cie Dowell, Sofitech NV; 2000. URL: https://www.google.at/patents/WO2000031154A1?cl=en.

[55] Barclay-Miller DJ, Martin DW, Wall K, Zard PW. Surfactant composition. WO Patent 1995030722 assigned to Burwood Corp. Ltd.; 1995. URL: https://www.google.at/patents/WO1995030722A1?cl=en.

[56] Brankling D. Drilling fluid. WO Patent 1994002565 assigned to David Brankling, Oilfield Chem Tech Ltd; 1994. URL: https://www.google.at/patents/WO1994002565A1?cl=en.

[57] Ujma KHW, Plank JP. A new calcium-tolerant polymer helps to improve drilling mud performance and reduce costs. In: Proceedings Volume. 62nd Annu. SPE Tech. Conf. (Dallas, 9/27-30/87); 1987, p. 327-34.

[58] Ujma KH, Sahr M, Plank J, Schoenlinner J. Cost reduction and improvement of drilling mud properties by using polydrill (Kostenreduzierung und Verbesserung der Spülungseigenschaften mit Polydrill). Erdöl Erdgas Kohle 1987;103(5):219-22.

[59] Plank J. Field results with a novel fluid loss polymer for drilling muds. Oil Gas Europe Mag 1990;16(3):20-3.

[60] Roark DN, Nugent Adam J, Bandlish BK. Fluid loss control and compositions for use therein. EP Patent 0201355 assigned to Ethyl Corporation; 1991. URL: https://www.google.at/patents/EP0201355B1?cl=en.

[61] Roark DN, Nugent Jr A, Bandlish BK. Fluid loss control in well cement slurries. US Patent 4698380, assigned to Ethyl Corporation (Richmond, VA); 1987. URL: http://www.freepatentsonline.com/4698380.html.

[62] Guichard B, Wood B, Vongphouthone P. Fluid loss reducer for high temperature high pressure water based-mud application. US Patent 7449430, assigned to Eliokem S.A.S. (Villejust, FR); 2008. URL: http://www.freepatentsonline.com/7449430.html.

[63] Patel BB. Drilling fluid additive and process therewith. US Patent 6124245, assigned to Phillips Petroleum Company (Bartlesville, OK); 2000. URL: http://www.freepatentsonline.com/6124245.html.

[64] Stephens M. Fluid loss additives for well cementing compositions. GB Patent 2202526; 1988.

[65] Stephens M, Swanson BL. Drilling mud comprising tetrapolymer consisting of n-vinyl-2-pyrrolidone, acrylamidopropanesulfonic acid, acrylamide, and acrylic acid. US Patent 5135909, assigned to Phillips Petroleum Company (Bartlesville, OK); 1992. URL: http://www.freepatentsonline.com/5135909.html.

[66] Arlt Dieter GDSC. Polymers containing sulphonic acid groups. US Patent 3547899, assigned to Bayer AG, Leverkusen; 1970. URL: http://www.freepatentsonline.com/3547899.html.

[67] Stephens M, Swanson BL, Patel BB. Drilling mud comprising tetrapolymer consisting of n-vinyl-2-pyrrolidone, acrylamidopropanesulfonicacid, acrylamide, and acrylic acid. US Patent 5380705, assigned to Phillips Petroleum Company (Bartlesville, OK); 1995. URL: http://www.freepatentsonline.com/5380705.html.

[68] Garvey CM, Savoly A, Resnick AL. Fluid loss control additives and drilling fluids containing same. US Patent 4741843, assigned to Diamond Shamrock Chemical (Dallas, TX); 1988. URL: http://www.freepatentsonline.com/4741843.html.

[69] Bardoliwalla DF. Fluid loss control additives from AMPS polymers. US Patent 4622373, assigned to Diamond Shamrock Chemicals Company (Dallas, TX); 1986. URL: http://www.freepatentsonline.com/4622373.html.

[70] Jean-Francois A, Annie AH, Lionel R. Filtrate reducing additive and well fluid. WO Patent 1998059014 assigned to Inst. Francais Du Petrole; 1998. URL: https://www.google.at/patents/WO1998059014A1?cl=en.

[71] Peiffer DG, Lundberg RD, Sedillo L, Newlove JC. Fluid loss control in oil field cements. US Patent 4626285, assigned to Exxon Research and Engineering Company (Florham Park, NJ); 1986. URL: http://www.freepatentsonline.com/4626285.html.

[72] Savoly A, Villa JL, Garvey CM, Resnick AL. Fluid loss agents for oil well cementing composition. US Patent 4674574, assigned to Diamond Shamrock Chemicals Company (Dallas, TX); 1987. URL: http://www.freepatentsonline.com/4674574.html.

[73] Zhang GP, Ye HC. AM/MA/AMPS terpolymer as non-viscosifying filtrate loss reducer for drilling fluids. Oilfield Chem 1998;15(3):269-71.

[74] Lange W, Bohmer B. Water-soluble polymers and their use as flushing liquid additives for drilling. US Patent 4749498, assigned to Wolff Walsrode Aktiengesellschaft (Walsrode, DE); 1988. URL: http://www.freepatentsonline.com/4749498.html.

[75] Jarrett M, Clapper D. High temperature filtration control using water based drilling fluid systems comprising water soluble polymers. US Patent 7651980, assigned to Baker Hughes Incorporated (Houston, TX); 2010. URL: http://www.freepatentsonline.com/7651980.html.

[76] Patel AD. Water based drilling fluids with high temperature fluid loss control additive. US Patent 5789349, assigned to M-I Drilling Fluids, L.L.C. (Houston, TX); 1998. URL: http://www.freepatentsonline.com/5789349.html.

[77] Ganguli KK. High temperature fluid loss additive for cement slurry and method of cementing. US Patent 5116421, assigned to The Western Company of North America (Houston, TX); 1992. URL: http://www.freepatentsonline.com/5116421.html.

[78] Brothers LE. Method of reducing fluid loss in cement compositions which may contain substantial salt concentrations. US Patent 4700780, assigned to Halliburton Services (Duncan, OK); 1987. URL: http://www.freepatentsonline.com/4700780.html.

[79] Crema SC, Kucera CH, Konrad G, Hartmann H. Fluid loss control additives for oil well cementing compositions. US Patent 5025040, assigned to BASF (Porsippany, NJ); 1991. URL: http://www.freepatentsonline.com/5025040.html.

[80] Crema SC, Kucera CH, Konrad G, Hartmann H. Fluid loss control additives for oil well cementing compositions. US Patent 5228915, assigned to BASF Corporation (Parsippany, NJ); 1993. URL: http://www.freepatentsonline.com/5228915.html.

[81] Kucera CH, Crema SC, Roznowski MD, Konrad G, Hartmann H. Fluid loss control additives for oil well cementing compositions. EP Patent 0342500 assigned to BASF Corporation; 1989. URL: https://www.google.at/patents/EP0342500A2?cl=en.

[82] Ghassemzadeh J. Lost circulation material for oilfield use. US Patent 8404622, assigned to Schlumberger Technology Corporation (Sugar Land, TX); 2013. URL: http://www.freepatentsonline.com/8404622.html.

[83] Hofstütter H. Bore hole fluid comprising dispersed synthetic polymeric fibers. WO Patent 2012123338 assigned to Lenzing Plastics Gmbh; 2012. URL: https://www.google.at/patents/WO2012123338A1?cl=en.

[84] Melbouci M, Sau AC. Water based drilling fluids. US Patent 7384892, assigned to Hercules Incorporated (Wilmington, DE); 2008. URL: http://www.freepatentsonline.com/7384892.html.

[85] Hen J. Sulfonate-containing polymer/polyanionic cellulose combination for high temperature/high pressure filtration control in water base drilling fluids. US Patent 5008025, assigned to Mobil Oil Corporation (Fairfax, VA); 1991. URL: http://www.freepatentsonline.com/5008025.html.

[86] Brothers LE. Method of reducing fluid loss in cement compositions containing substantial salt concentrations. US Patent 4640942, assigned to Halliburton Company (Duncan, OK); 1987. URL: http://www.freepatentsonline.com/4640942.html.

[87] Raines RH. Use of low m.s. hydroxyethyl cellulose for fluid loss control in oil well applications. US Patent 4629573, assigned to Union Carbide Corporation (Danbury, CT); 1986. URL: http://www.freepatentsonline.com/4629573.html.

[88] Chang FF, Bowman M, Parlar M, Ali SA, Cromb J. Development of a new crosslinked-hec (hydroxyethylcellulose) fluid loss control pill for highly-overbalanced, high-permeability and/or high temperature formations. In:

Proceedings Volume. SPE Formation Damage Contr. Int. Symp. (Lafayette, LA, 2/18-19/98); 1998, p. 215-27.

[89]   Chang FF, Parlar M. Method and composition for controlling fluid loss in high permeability hydrocarbon bearing formations. US Patent 5981447, assigned to Schlumberger Technology Corporation (Sugar Land, TX); 1999. URL: http://www.freepatentsonline.com/5981447.html.

[90]   Nguyen PD, Weaver JD, Cole RC, Schulze CR. Development and field application of a new fluid-loss control material. In: Proceedings Volume. Annu. SPE Tech. Conf. (Denver, 10/6-9/96); 1996, p. 933-41. URL: http://www.onepetro.org/mslib/servlet/onepetropreview?id=00036676. doi:10.2118/36676-MS.

[91]   Audibert A, Argillier JF, Bailey L, Reid PI. Process and water-base fluid utilizing hydrophobically modified cellulose derivatives as filtrate reducers. US Patent 5669456, assigned to Institut Francais du Petrole (Rueil-Malmaison, FR) Dowell Schlumberger, Inc. (Sugar Land, TX); 1997. URL: http://www.freepatentsonline.com/5669456.html.

[92]   Amanullah M, Yu L. Environment friendly fluid loss additives to protect the marine environment from the detrimental effect of mud additives. J Pet Sci Eng 2005;48 (3-4):199-208.

[93]   Francis HP, DeBoer ED, Wermers VL. High temperature drilling fluid component. US Patent 4652384, assigned to American Maize-Products Company (Hammond, IN); 1987. URL: http://www.freepatentsonline.com/4652384.html.

[94]   Nguyen N, Sifferman TR, Skaggs CB, Solarek DB, Swazey JM. Fluid loss control additives and subterranean treatment fluids containing the same. WO Patent 1999005235 assigned to Monsanto Co., Nat. Starch Chem. Invest.; 1999. URL: https://www.google.at/patents/WO1999005235A1?cl=en.

[95]   API Standard RP 13I. Recommended practice for laboratory testing of drilling fluids. API Standard API RP 13I; American Petroleum Institute; Washington, DC; 2009.

[96]   ASTM D5891-02. Standard test method for fluid loss of clay component of geosynthetic clay liners. ASTM Standard, Book of Standards, Vol. 04.13 ASTM D5891-02; ASTM International; West Conshohocken, PA; 2009.

[97]   Zhang CG, Sun MB, Hou WG, Sun D. Study on function mechanism of filtration reducer: The influence of fluid loss additive on electrical charge density of filter cake fines. Drilling Fluid and Completion Fluid 1995;12(4):1-5.

[98]   Zhang CG, Sun MB, Hou WG, Liu YY, Sun DJ. Study on function mechanism of filtration reducer: Comparison. Drilling Fluid and Completion Fluid 1996;13(3): 11-17.

[99]   Plank JP, Gossen FA. Visualization of fluid-loss polymers in drilling mud filter cakes. In: Proceedings Volume. 64th Annu. SPE Tech. Conf. (San Antonio, 10/8-11/89); 1989, p. 165-76.

[100]   Ballerini D, Choplin L, Dreveton E, Lecourtier J. Method using gellan for reducing the filtrate of aqueous drilling fluids. EP Patent 0662563 assigned to Institut Francais Du Petrole; 2000. URL: https://www.google.at/patents/EP0662563B1?cl=en.

[101]   Dreveton E, Lecourtier J, Ballerini D, Choplin L. Process using gellan as a filtrate reducer for water based drilling fluids. US Patent 5744428, assigned to Institut Francais Du Petrole (Rueil-Malmaison, FR); 1998. URL: http://www.freepatentsonline.com/5744428.html.

[102]   Navarrete RC, Seheult JM, Himes RE. Applications of xanthan gum in fluid-loss control and related formation damage. In: Proceedings Volume. SPE Permian Basin Oil & Gas Recovery Conf. (Midland, TX, 3/21-23/2000); 2000.

[103]   Bruno L. Fluids useful for oil mining comprising de-acetylated xanthane gum and at least one compound increasing the medium ionic strength. WO Patent 1999003948 assigned to Rhone Poulenc Chimie; 1999. URL: https://www.google.at/patents/WO1999003948A1?cl=en.

[104]   Wikipedia,. Leonardite – wikipedia, the free encyclopedia; 2013. URL: http://en. wikipedia.org/w/index.php?title=Leonardite&oldid=530716648; [Online; accessed 19-February-2014].

[105]   Hayes J. Dry mix for water based drilling fluid. US Patent 6818596; 2004. URL: http://www.freepatentsonline.com/6818596.html.

[106]   Tan D. Test and application of drilling fluid filtrate reducer polysulfonated humic acid resin. Oil Drilling Prod Technol 1990;12(1):27-32,97-98.

[107]   Firth Jr WC. Chrome humates as drilling mud additives. US Patent 4921620, assigned to Union Camp Corporation (Wayne, NJ); 1990. URL: http://www. freepatentsonline.com/4921620.html.

[108]   Xiang T. Drilling fluid systems for reducing circulation losses. US Patent 7226895, assigned to Baker Hughes Incorporated (Houston, TX); 2007. URL: http://www. freepatentsonline.com/7226895.html.

[109]   Xiang T. Methods for reducing circulation loss during drilling operations. US Patent 7507692, assigned to Baker Hughes Incorporated (Houston, TX); 2009. URL: http://www.freepatentsonline.com/7507692.html.

[110]   Brand FJ, Bradbury A. Water based wellbore fluids. US Patent 6884760, assigned to M-I, L.L.C. (Houston, TX); 2005. URL: http://www.freepatentsonline.com/ 6884760.html.

[111]   Dobson Jr JW, Kayga PD. Magnesium peroxide breaker system improves filter cake removal. Pet Eng Int 1995;68(10):49-50.

[112]   Al Moajil AM, Nasr-El-Din HA. Removal of manganese tetraoxide filter cake using a combination of HCl and organic acid. Journal of Canadian Petroleum Technology 2014;53(02):122-30. doi:10.2118/165551-pa.

[113]   Reddy BR, Palmer AV. Sealant compositions comprising colloidally stabilized latex and methods of using the same. US Patent 7607483, assigned to Halliburton Energy Services, Inc. (Duncan, OK); 2009. URL: http://www.freepatentsonline.com/ 7607483.html.

[114]   Stowe II CJ, Bland RG, Clapper DK, Xiang T, Benaissa S. Water based drilling fluids using latex additives. US Patent 6703351, assigned to Baker Hughes Incorporated (Houston, TX); 2004. URL: http://www.freepatentsonline.com/6703351.html.

[115]   Halliday WS, Schwertner D, Xiang T, Clapper DK. Water based drilling fluids using latex additives. US Patent 7393813, assigned to Baker Hughes Incorporated (Houston, TX); 2008. URL: http://www.freepatentsonline.com/7393813.html.

[116]   Halliday WS, Schwertner D, Xiang T, Clapper DK. Fluid loss control and sealing agent for drilling depleted sand formations. US Patent 7271131, assigned to Baker Hughes Incorporated (Houston, TX); 2007. URL: http://www.freepatentsonline. com/7271131.html.

[117]   Bailey L. Latex additive for water based drilling fluids. US Patent 6715568, assigned to M-I L.L.C. (Houston, TX); 2004. URL: http://www.freepatentsonline.com/ 6715568.html.

[118]   Hernandez MI, Mas M, Gabay RJ, Quintero L. Thermally stable drilling fluid. US Patent 5883054, assigned to Intevep, S.A. (Caracas 1070A, VE); 1999. URL: http://www.freepatentsonline.com/5883054.html.

[119]   Sweatman RE, Felio AJ, Heathman JF. Water based compositions for sealing subterranean zones and methods. US Patent 6258757, assigned to Halliburton Energy Services, Inc. (Duncan, OK); 2001. URL: http://www.freepatentsonline. com/6258757.html.

[120]   Zhou M, Qiu X, Yang D, Wang W. Synthesis and evaluation of sulphonated acetone formaldehyde resin applied as dispersant of coal water slurry. Energy Conversion and Management 2007;48(1):204-9. URL: http://www. sciencedirect.com/science/article/pii/S019689040600152X.   http://dx.doi.org /10.1016/j.enconman.2006.04.015.

[121] Liu M, Lei J, Du X, Fu C, Huang B. Synthesis, properties and dispersion mechanism of sulphonated acetone-formaldehyde superplasticizer in cementitious system. J Wuhan Univ Technol Mater Sci Ed 2013;28(6):1167-71. doi:10.1007/s11595-013-0838-7.

[122] Rayborn Sr JJ, Dickerson JP. Method of making a drilling fluid containing carbon black in a dispersed state. US Patent 5114597, assigned to Sun Drilling Products Corporation (Belle Chasse, LA); 1992. URL: http://www.freepatentsonline.com/5114597.html.

[123] Rayborn Sr JJ, Rayborn JJ. Drilling fluid system containing a combination of hydrophilic carbon black/asphaltite and a refined fish oil/glycol mixture and related methods. US Patent 5942467, assigned to Sun Drilling Products Corporation (Belle Chasse, LA); 1999. URL: http://www.freepatentsonline.com/5942467.html.

[124] Vitthal S, McGowen JM. Fracturing fluid leakoff under dynamic conditions: Pt.2: Effect of shear rate, permeability, and pressure. In: Proceedings Volume. Annu. SPE Tech. Conf. (Denver, 10/6-9/96); 1996, p. 821-35.

[125] Lord DL, Vinod PS, Shah S, Bishop ML. An investigation of fluid leakoff phenomena employing a high-pressure simulator. In: Proceedings Volume. Annu. SPE Tech. Conf. (Dallas, 10/22-25/95); 1995, p. 465-74.

[126] Buret S, Nabzar L, Jada A. Water quality and well injectivity: do residual oil-in-water emulsions matter? SPE J 2010;15(2):557-68. doi:10.2118/122060-pa.

[127] Müller H, Herzog N, Behler A, Hartmann J. Borehole treating substance containing ether carboxylic acids. US Patent 7741248, assigned to Emery Oleochemicals GmbH (Dusseldorf, DE); 2010. URL: http://www.freepatentsonline.com/7741248.html.

[128] Klug P, Kupfer R, Wimmer I, Winter R. Process for the preparation of ether carboxylic acids with low residual alcohol content. US Patent 6326514, assigned to Clariant GmbH (Frankfurt, DE); 2001. URL: http://www.freepatentsonline.com/6326514.html.

[129] Thaemlitz CJ. Synthetic filtration control polymers for wellbore fluids. US Patent 7098171, assigned to Halliburton Energy Services, Inc. (Duncan, OK); 2006. URL: http://www.freepatentsonline.com/7098171.html.

[130] Halliday WS, Clapper DK, Smalling MR. Glycols as gas hydrate inhibitors in drilling, drill-in, and completion fluids. US Patent 6080704; 2000. URL: http://www.freepatentsonline.com/6080704.html.

[131] Lugo R, Dalmazzone C, Audibert A. Method and thermodynamic inhibitors of gas hydrates in water based fluids. US Patent 7709419, assigned to Institut Francais du Petrole (Rueil Malmaison Cedex, FR); 2010. URL: http://www.freepatentsonline.com/7709419.html.

[132] EPA Toxicity Test. Mysid acute toxicity test. Toxicity Test EPA 712-C-96-136; US Environmental Protection Agency; Washington, DC; 1996. URL: http://www.epa.gov/ocspp/pubs/frs/publications/OPPTS_Harmonized/850_Ecological_Effects_Test_Guidelines/Drafts/850-1035.pdf.

[133] Kotkoskie TS, Al-Ubaidi B, Wildeman TR, Sloan ED. Inhibition of gas hydrates in water based drilling muds. SPE Drilling Engineering 1992;7(2). doi:10.2118/20437-PA.

[134] Knox D, Jiang P. Drilling further with water based fluids – selecting the right lubricant. In: Proceedings Volume. 92002-MS; International Symposium on Oilfield Chemistry; The Woodlands, Texas: Society of Petroleum Engineers, Inc.; 2005.

[135] Halliday WS, Schwertner D. Olefin isomers as lubricants, rate of penetration enhancers, and spotting fluid additives for water based drilling fluids. US Patent 5605879, assigned to Baker Hughes Incorporated (Houston, TX); 1997. URL: http://www.freepatentsonline.com/5605879.html.

[136]   Westfechtel A, Maker D, Muller H. Oligoglyercol fatty acid ester additives for water based drilling fluids. US Patent 8148305, assigned to Emery Oleochemicals GmbH (Dusseldorf, DE); 2012. URL: http://www.freepatentsonline.com/8148305.html.

[137]   Chapman J, Ward I. Lubricant for drilling mud. EP Patent 0770661 assigned to B W Mud Limited; 1997. URL: https://www.google.at/patents/EP0770661A1?cl=en.

[138]   Müller H, Herold CP, Bongardt F, Herzog N, von Tapavicza S. Lubricants for drilling fluids. US Patent 6806235, assigned to Cognis Deutschland GmbH & Co. KG (Düsseldorf, DE); 2004. URL: http://www.freepatentsonline.com/6806235. html.

[139]   Breeden DL, Meyer RL. Ester-containing downhole drilling lubricating composition and processes therefor and therewith. US Patent 6884762, assigned to Newpark Drilling Fluids, L.L.C. (Houston, TX); 2005. URL: http://www. freepatentsonline.com/6884762.html.

[140]   Durr Albert MJ, Huycke J, Jackson HL, Hardy BJ, Smith KW. An ester base oil for lubricant compounds and process of making an ester base oil from an organic reaction by-product. EP Patent 0606553 assigned to Conoco Inc.; 1994. URL: https://www.google.at/patents/EP0606553A2?cl=en.

[141]   Genuyt B, Janssen M, Reguerre R, Cassiers J, Breye F. Biodegradable lubricating composition and uses thereof, in particular in a bore fluid [composition lubrifiante biodegradable et ses utilisations, notamment dans un fluide de forage]. WO Patent 0183640, assigned to Total Raffinage Dist SA; 2001.

[142]   Senaratne KPA, Lilje KC. Preparation of branched chain carboxylic esters. US Patent 5322633, assigned to Albemarle Corporation (Richmond, VA); 1994. URL: http://www.freepatentsonline.com/5322633.html.

[143]   Runov VA, Mojsa YN, Subbotina TV, Pak KS, Krezub AP, Pavlychev VN, et al. Lubricating additive for clayey drilling solution – is obtained by esterification of tall oil or tall pitch with hydroxyl group containing agent, e.g. low mol. wt. glycol or ethyl cellulose. SU Patent 1700044, assigned to Volgo Don Br. Sintez Pav and Burenie Sci Prod. Assoc.; 1991.

[144]   Andreson BA, Abdrakhmanov RG, Bochkarev GP, Umutbaev VN, Fryazinov VV, Kudinov VN, et al. Lubricating additive for water based drilling solutions – contains products of condensation of monoethanolamine and tall oils, kerosene, monoethanolamine and flotation reagent. SU Patent 1749226, assigned to Bashkir Oil Ind. Res. Inst. and Bashkir Oil Proc. Inst.; 1992.

[145]   Argillier JF, Demoulin A, Audibert-Hayet A, Janssen M. Borehole fluid containing a lubricating composition – method for verifying the lubrification of a borehole fluid – application with respect to fluids with a high ph (fluide de puits comportant une composition lubrifiante – procede pour controler la lubrification d'un fluide de puits – application aux fluides a haut ph). WO Patent 9966006, assigned to Inst. Francais Du Petrole and Fina Research SA; 1999.

[146]   Kashkarov NG, Verkhovskaya NN, Ryabokon AA, Gnoevykh AN, Konovalov EA, Vyakhirev VI. Lubricating reagent for drilling fluids – consists of spent sunflower oil modified with additive in form of aqueous solutions of sodium alkylsiliconate(s). RU Patent 2076132, assigned to Tyumen Nat Gases Res. Inst.; 1997.

[147]   Kashkarov NG, Konovalov EA, Vjakhirev VI, Gnoevykh AN, Rjabokon AA, Verkhovskaja NN. Lubricant reagent for drilling muds – contains spent sunflower oil, and light tall oil and spent coolant-lubricant as modifiers. RU Patent 2105783, assigned to Tyumen Nat Gases Res. Inst.; 1998.

[148]   Konovalov EA, Ivanov YA, Shumilina TN, Pichugin VF, Komarova NN. Lubricating reagent for drilling solutions – contains agent based on spent sunflower oil, water, vat residue from production of oleic acid, and additionally water glass. SU Patent 1808861, assigned to Moscow Gubkin Oil Gas Inst.; 1993.

[149]　Konovalov EA, Rozov AL, Zakharov AP, Ivanov YA, Pichugin VF, Komarova NN. Lubricating reagent for drilling solutions – contains spent sunflower oil as active component, water, boric acid as emulsifier, and additionally water glass. SU Patent 1808862, assigned to Moscow Gubkin Oil Gas Inst.; 1993.

[150]　Bel SLA, Demin VV, Kashkarov NG, Konovalov EA, Sidorov VM, Bezsolitsen VP, et al. Lubricating composition – for treatment of clayey drilling solutions, contains additive in form of sulphonated fish fat. RU Patent 2106381, assigned to Shchelkovsk Agro Ent St C and Fakel Res. Prod. Assoc.; 1998.

[151]　Müller H, Herold CP, Bongardt F, Herzog N, von Tapavicza S. Lubricants for drilling fluids (Schmiermittel für Bohrspülungen). WO Patent 0029502, assigned to Cognis Deutschland GmbH; 2000.

[152]　Wall K, Martin DW, Zard PW, Barclay-Miller DJ. Temperature stable synthetic oil. WO Patent 9532265, assigned to Burwood Corp. Ltd.; 1995.

[153]　Argillier JF, Audibert A, Marchand P, Demoulin Ae, Janssen M. Lubricating composition including an ester-use of the composition and well fluid including the composition. US Patent 5618780, assigned to Institut Francais Du Petrole (Rueil-Malmaison, FR); 1997. URL: http://www.freepatentsonline.com/5618780.html.

[154]　Müller H, Herold CP, von Tapavicza S. Use of selected fatty alcohols and their mixtures with carboxylic acid esters as lubricant components in water based drilling fluid systems for soil exploration. US Patent 6716799, assigned to Cognis Deutschland GmbH & Co. KG (Düsseldorf, DE); 2004. URL: http://www.freepatentsonline.com/6716799.html.

[155]　Malchow Jr GA. Water based drilling fluids containing phosphites as lubricating aids. US Patent 5807811, assigned to The Lubrizol Corporation (Wickliffe, OH); 1998. URL: http://www.freepatentsonline.com/5807811.html.

[156]　Dixon J. Drilling fluids. US Patent 7614462, assigned to Croda International PLC (Goole, East Yorkshire, GB); 2009. URL: http://www.freepatentsonline.com/7614462.html.

[157]　Dixon J. Drilling fluids. US Patent 7343986, assigned to Croda International PLC (Goole, East Yorkshire, GB); 2008. URL: http://www.freepatentsonline.com/7343986.html.

[158]　Malchow Jr GA. Friction modifier for water based well drilling fluids and methods of using the same. US Patent 5593954, assigned to The Lubrizol Corporation (Wickliffe, OH); 1997. URL: http://www.freepatentsonline.com/5593954.html.

[159]　Wu A, Yan X. Silicate drilling fluid composition containing lubricating agents and uses thereof. US Patent 7842651, assigned to Chengdu Cationic Chemistry Company, Inc. (Sichuan, CN); 2010. URL: http://www.freepatentsonline.com/7842651.html.

[160]　Vail JG, Baker LC. Compositions for and process of making suspensions. US Patent 2133759, assigned to Philadelphia Quartz Company; 1938. URL: http://www.freepatentsonline.com/2133759.html.

[161]　Dearing Jr HL. Silicate drilling fluid and method of drilling a well therewith. US Patent 7137459, assigned to Newpark Drilling Fluids (Houston, TX); 2006. URL: http://www.freepatentsonline.com/7137459.html.

[162]　Bailey L. Water based drilling fluid. US Patent 7833946, assigned to M-I, L.L.C. (Houston, TX); 2010. URL: http://www.freepatentsonline.com/7833946.html.

[163]　Huber J, Plank J, Heidlas J, Keilhofer G, Lange P. Additive for drilling fluids. US Patent 7576039, assigned to BASF Construction Polymers GmbH (Trostberg, DE); 2009. URL: http://www.freepatentsonline.com/7576039.html.

[164]　Rayborn Sr JJ. Water based drilling fluid additive containing graphite and carrier. US Patent 7067461, assigned to Alpine Mud Products Corp. (Belle Chasse, LA); 2006. URL: http://www.freepatentsonline.com/7067461.html.

[165]   Halliday WS, Clapper DK. Purified paraffins as lubricants, rate of penetration enhancers, and spotting fluid additives for water based drilling fluids. US Patent 5837655; 1998. URL: http://www.freepatentsonline.com/5837655.html.

[166]   Buyanovskii IA. Tribological test methods and apparatus. Chem Technol Fuels Oils 1994;30(3):133-47.

[167]   ASTM D 3233(93). Standard test methods for measurement of extreme pressure properties of fluid lubricants (falex pin and vee block methods). ASTM Standard, Book of Standards, Vol. 5.01 ASTM D 3233(93); ASTM International; West Conshohocken, PA; 2009.

[168]   ASTM D 2782-02. Standard test method for measurement of extreme-pressure properties of lubricating fluids (timken method). ASTM Standard, Book of Standards, Vol. 5.01 ASTM D 2782-02; ASTM International; West Conshohocken, PA; 2008.

[169]   Totten GE, Westbrook SR, Shah RJ, editors. Fuels and lubricants handbook: technology, properties, performance, and testing; ASTM Manual Series vol. 37. West Conshohocken, PA: American Society for Testing & Materials (ASTM); 2003.

[170]   Argillier JF, Demoulin A, Audibert-Hayet A, Janssen M. Borehole fluid containing a lubricating composition-method for verifying the lubrification of a borehole fluid–application with respect to fluids with a high ph. US Patent 6750180, assigned to Institut Francais du Petrole (Rueil-Malmaison Cedex, FR) Oleon NV (Ertvelde, BE); 2004. URL: http://www.freepatentsonline.com/6750180.html.

[171]   Alonso-DeBolt MA, Bland RG, Chai BJ, Eichelberger PB, Elphingstone EA. Glycol and glycol ether lubricants and spotting fluids. US Patent 5945386, assigned to Baker Hughes Incorporated (Houston, TX); 1999. URL: http://www.freepatentsonline.com/5945386.html.

[172]   Penkov AI, Vakhrushev LP, Belenko EV. Characteristics of the behavior and use of polyalkylene glycols for chemical treatment of drilling muds. Stroit Neft Gaz Skvazhin Sushe More 1999;(1-2):21-4.

[173]   Sano M. Polypropylene glycol (PPG) used as drilling fluids additive. Sekiyu Gakkaishi 1997;40(6):534-8.

[174]   Audebert R, Janca J, Maroy P, Hendriks H. New, chemically crosslinked polyvinyl alcohol (pva), process for synthesizing same and its applications as a fluid loss control agent in oil fluids. CA Patent 2118070 assigned to Roland Audebert, Joseph Janca, Pierre Maroy, Hugo Hendriks, Schlumberger Canada Limited; 2008. URL: https://www.google.at/patents/CA2118070C?cl=en.

[175]   Audebert R, Hendriks H, Janca J, Maroy P. Chemically crosslinked polyvinyl alcohol (pva), and its applications as a fluid loss control agent in oil fluids. EP Patent 0705850 assigned to Sofitech N.V.; 1998. URL: https://www.google.at/patents/EP0705850B1?cl=en.

[176]   Forsberg JW, Jahnke RW. Methods of drilling well boreholes and compositions used therein. US Patent 5260268, assigned to The Lubrizol Corporation (Wickliffe, OH); 1993. URL: http://www.freepatentsonline.com/5260268.html.

[177]   Forsberg JW, Jahnke RW. Methods of drilling well boreholes and compositions used therein. WO Patent 1993002151 assigned to Lubrizol Corp.; 1993. URL: https://www.google.at/patents/WO1993002151A1?cl=en.

[178]   Naraghi AR, Rozell RS. Method for reducing torque in downhole drilling. US Patent 5535834, assigned to Champion Technologies, Inc. (Fresno, TX); 1996. URL: http://www.freepatentsonline.com/5535834.html.

[179]   Clark DE, Dye WM. Environmentally safe lubricated well fluid method of making a well fluid and method of drilling. US Patent 5658860, assigned to Baker Hughes Incorporated (Houston, TX); 1997. URL: http://www.freepatentsonline.com/5658860.html.

[180] Dye W, Clark DE, Bland RG. Well fluid additive. EP Patent 0652271 assigned to Baker-Hughes Incorporated; 1995. URL: https://www.google.at/patents/EP0652271A1?cl=en.

[181] Herold CP, Müller H, Von Tapavicza S. Use of selected fatty alcohols and their mixtures with carboxylic acid esters as lubricant components in water based drilling fluid systems for soil exploration. EP Patent 0948576 assigned to Cognis Deutschland GmbH; 2001. URL: https://www.google.at/patents/EP0948576B1?cl=en.

[182] Xiang T, Amin RAM. Water based mud lubricant using fatty acid polyamine salts and fatty acid esters. US Patent 8413745, assigned to Baker Hughes Incorporated (Houston, TX); 2013. URL: http://www.freepatentsonline.com/8413745.html.

[183] Müller H., Herold C.P.. Powder-form lubricant additives for water based drilling fluids. US Patent 4802998, assigned to Henkel Kommanditgesellschaft auf Aktien (Düsseldorf, DE); 1989. URL: http://www.freepatentsonline.com/4802998.html.

[184] Reichert H. Lubricant Measuring Machine. DE Patent 1749247 assigned to Hermann Reichert; 1957. URL: https://www.google.at/patents/DE1749247U?cl=en.

[185] Illi W. Bestimmung des Druckaufnahmevermögens mit der Reibverschleißwaage nach Reichert. Arbeitsblatt 6; Verbraucherkreis – Industrieschmierstoffe; Daimler AG, Stuttgart (DE); 2005. URL: http://www.vkis.org/051201/VKIS %20Arbeitsblatt%206.pdf.

[186] Herold CP, Müller H, Foerster T, Von Tapavicza S, Claas M. Multiphase lubricant concentrates for use in water based systems in the field of exploratory soil drilling. US Patent 6211119, assigned to Henkel Kommanditgesellschaft auf Aktien (Düsseldorf, DE); 2001. URL: http://www.freepatentsonline.com/6211119.html.

[187] Adams EK. Downhole well lubricant. US Patent 5700767, assigned to CJD Investments, Inc. (Corpus Christi, TX); 1997. URL: http://www.freepatentsonline.com/5700767.html.

[188] Kalb R, Hofstätter H. Method of treating a borehole and drilling fluid. US Patent Application 20120103614, assigned to Montanuniversitat Leoben, Leoben (AT); 2012. URL: http://www.freepatentsonline.com/20120103614.html.

[189] Wasserscheid P. Ionic Liquids in Synthesis. Weinheim: Wiley-VCH; 2008. ISBN 9783527312399.

[190] Maghrabi S, Kulkarni D, Teke K, Kulkarni SD, Jamison D. Modeling of shale-erosion behavior in aqueous drilling fluids. In: Technical Session 8: Well Construction 3. SPE/EAGE European Unconventional Resources Conference and Exhibition; EAGE; 2014, URL: http://www.earthdoc.org/publication/publicationdetails/?publication=75108.

[191] Carbajal DL, Shumway W, Ezell RG. Inhibitive water based drilling fluid system and method for drilling sands and other water-sensitive formations. US Patent 7825072, assigned to Halliburton Energy Services Inc. (Duncan, OK); 2010. URL: http://www.freepatentsonline.com/7825072.html.

[192] Su J, Chu Q, Ren M. Properties of high temperature resistance and salt tolerance drilling fluids incorporating acrylamide/2-acrylamido-2-methyl-1-propane sulfonic acid/N-vinylpyrrolidone/dimethyl diallyl ammonium chloride quadripolymer as fluid loss additives. J Polymer Eng 2014;34(2):153-9. doi:10.1515/polyeng-2013-0270.

[193] Patel AD, Stamatakis E, Young S. High performance water based drilling mud and method of use. US Patent 7497262, assigned to M-I L.L.C. (Houston, TX); 2009. URL: http://www.freepatentsonline.com/7497262.html.

[194] Patel AD, Stamatakis E, Young S. High performance water based drilling mud and method of use. US Patent 7514389, assigned to M-I L.L.C. (Houston, TX); 2009. URL: http://www.freepatentsonline.com/7514389.html.

[195] Zhong HY, Qiu ZS, Huang WA, Qiao J, Li HB, Cao J. The development and application of a novel polyamine water-based drilling fluid. Petroleum Sci Tech 2014;32(4):497–504. doi:10.1080/10916466.2011.592897.

[196] Patel AD, Stamatakis E. Low conductivity water based wellbore fluid. US Patent 8598095, assigned to M-I L.L.C. (Houston, TX); 2013. URL: http://www.freepatentsonline.com/8598095.html.

[197] Patel AD. Low toxicity shale hydration inhibition agent and method of use. US Patent 8298996, assigned to M-I L.L.C. (Houston, TX); 2012. URL: http://www.freepatentsonline.com/8298996.html.

[198] Zhong HY, Qiu ZS, Huang WA, Cao J, Wang FW, Xie BQ. Inhibition comparison between polyether diamine and quaternary ammonium salt as shale inhibitor in water based drilling fluid. Energy Sources, Part A: Recovery, Utilization, and Environmental Effects 2013;35(3):218–25. URL: http://www.tandfonline.com/doi/abs/10.1080/15567036.2011.606871. doi:10.1080/15567036.2011.606871.

[199] Miller RF. Shale hydration inhibition agent(s) and method of use. US Patent 8026198, assigned to Shrieve Chemical Products, Inc.; 2011. URL: http://www.freepatentsonline.com/8026198.html.

[200] Smith CK, Balson TG. Shale-stabilizing additives. US Patent 6706667; 2004. URL: http://www.freepatentsonline.com/6706667.html.

[201] Kubena Jr E, Whitebay LE, Wingrave JA. Method for stabilizing boreholes. US Patent 5211250, assigned to Conoco Inc. (Ponca City, OK); 1993. URL: http://www.freepatentsonline.com/5211250.html.

[202] Dérand H, Wesslén B, Wittgren B, Wahlund KG. Poly (ethylene glycol) graft copolymers containing carboxylic acid groups: Aggregation and viscometric properties in aqueous solution. Macromolecules 1996;29(27):8770–75.

[203] Carminati S, Brignoli M. Water based drilling fluid containing anions with a high hydrodynamic radius. US Patent 6500785, assigned to ENI S.p.A. (Rome, IT) Enitecnologie S.p.A. (San Donato Milanese, IT); 2002. URL: http://www.freepatentsonline.com/6500785.html.

[204] Carminati S, Guarneri A, Brignoli M. Drilling fluids comprising oil emulsions in water. US Patent 6632777, assigned to ENI S.p.A. (Rome, IT) Enitecnologie S.p.A (San Donato, IT); 2003. URL: http://www.freepatentsonline.com/6632777.html.

[205] Korzilius J, Minks P. Alkali-metal-carboxylate-containing drilling fluid having improved corrosion properties. US Patent 6239081, assigned to Clariant GmbH (Frankfurt, DE); 2001. URL: http://www.freepatentsonline.com/6239081.html.

[206] Miksic BA, Furman A, Kharshan M, Braaten J, Leth-Olsen H. Corrosion resistant system for performance drilling fluids utilizing formate brine. US Patent 6695897, assigned to Cortec Corporation (St. Paul, MN); 2004. URL: http://www.freepatentsonline.com/6695897.html.

[207] Wu Y. Corrosion inhibitor for wellbore applications. US Patent 5654260, assigned to Phillips Petroleum Company (Bartlesville, OK); 1997. URL: http://www.freepatentsonline.com/5654260.html.

[208] Enright D, Dye W, Smith M. An environmentally safe water based alternative to oil muds. SPE Drilling Eng 1992;7(1). doi:10.2118/21937-PA.

[209] Zijsling DH, Illerhaus R. Eggbeater PDC drillbit design eliminates balling in water based drilling fluids. SPE Drill Complet 1993;8(4). doi:10.2118/21933-PA.

[210] Bailey L, Grover B. Anti-accretion additives for drilling fluids. US Patent 6803346, assigned to Schlumberger Technology Corporation (Sugar Land, TX); 2004. URL: http://www.freepatentsonline.com/6803346.html.

[211] Patel AD. Aqueous based drilling fluid additive and composition. US Patent 5639715, assigned to M-I Drilling Fluids LLC; 1997. URL: http://www.freepatentsonline.com/5639715.html.

[212] Lynn JD, Nasr-El-Din HA. Formation Damage Associated with water based Drilling Fluids and Emulsified Acid Study. Society of Petroleum Engineers. ISBN 9781555633615; 1999, doi:10.2118/54718-MS.

[213] Bishop S. The Experimental Investigation of Formation Damage Due to the Induced Flocculation of Clays Within a Sandstone Pore Structure by a High Salinity Brine. Society of Petroleum Engineers. ISBN 9781555634056; 1997, doi:10.2118/38156-MS.

[214] Al-Yami A, Nasr-El-Din H, Al-Shafei M, Bataweel M. Impact of water based drilling-in fluids on solids invasion and damage characteristics. SPE Product Operat 2010;25(1). doi:10.2118/117162-PA.

[215] Bailey L, Boek E, Jacques S, Boassen T, Selle O, Argillier JF, et al. Particulate Invasion From Drilling Fluids. Society of Petroleum Engineers. ISBN 9781555633615; 1999, doi:10.2118/54762-MS.

[216] Zain Z, Sharma M. Mechanisms of mudcake removal during flowback. SPE Drill Complet 2001;16(4). doi:10.2118/74972-PA.

[217] Moajil AMA, Nasr-El-Din HA, Al-Yami AS, Al-Aamri AD, Al-Agil AK. Removal of filter cake formed by manganese tetraoxide-based drilling fluids. In: SPE International Symposium and Exhibition on Formation Damage Control. Lafayette, Louisiana, USA: Society of Petroleum Engineers; 2008.

[218] Al-Yami ASHAB. Non-damaging manganese tetroxide water based drilling fluids. US Patent 7618924, assigned to Saudi Arabian Oil Company (Dhahran, SA); 2009. URL: http://www.freepatentsonline.com/7618924.html.

[219] Al-Yami ASHAB. Non-damaging manganese tetroxide water based drilling fluids. US Patent 7732379, assigned to Saudi Arabian Oil Company (Dhahran, SA); 2010. URL: http://www.freepatentsonline.com/7732379.html.

[220] Quintero L. Surfactant-polymer composition for substantially solid-free water based drilling, drill-in, and completion fluids. US Patent 7148183, assigned to Baker Hughes Incorporated (Houston, TX); 2006. URL: http://www.freepatentsonline.com/7148183.html.

[221] Neel KR. Gilsonite. In: Kirk-Othmer Encyclopedia of Chemical Technology; vol. 11; 3 ed. New York: J. Wiley & Sons; 1980, p. 802-6.

[222] Christensen KC, Davis Neal I, Nuzzolo M. Water-wettable drilling mud additives containing uintaite. US Patent RE35163 assigned to American Gilsonite Comp.; 1996. URL: http://www.freepatentsonline.com/RE35163.html.

[223] Christensen KC, Davis IN, Nuzzolo M. Water-wettable drilling mud additives containing uintaite. EP Patent 0460067 assigned to American Gilsonite Company; 1996. URL: https://www.google.at/patents/EP0460067B1?cl=en.

[224] Christensen KC, Davis IN, Nuzzolo M. Water-wettable drilling mud additives containing uintaite. WO Patent 1990010043 assigned to Chevron Res. & Tech.; 1990. URL: https://www.google.at/patents/WO1990010043A1?cl=en.

[225] Christensen KC, Davis II N, Nuzzolo M. Water-wettable drilling mud additives containing uintaite. US Patent 5030365, assigned to Chevron Research Company (San Francisco, CA); 1991. URL: http://www.freepatentsonline.com/5030365.html.

[226] Rayborn Sr JJ, Dickerson JP. Method of making drilling fluid containing asphaltite in a dispersed state. US Patent 5114598, assigned to Sun Drilling Products Corporation (Belle Chasse, LA); 1992. URL: http://www.freepatentsonline.com/5114598.html.

[227] Patel BB. Liquid additive comprising a sulfonated asphalt and processes therefor and therewith. US Patent 5502030, assigned to Phillips Petroleum Company (Bartlesville, OK); 1996. URL: http://www.freepatentsonline.com/5502030.html.

[228] Anderson RL, Ratcliffe I, Greenwell HC, Williams PA, Cliffe S, Coveney PV. Clay swelling – A challenge in the oilfield. Earth-Sci Rev 2010;98(3-4): 201-16.

[229] Mooney RW, Keenan AG, Wood LA. Adsorption of water vapor by montmorillonite. II. Effect of exchangeable ions and lattice swelling as measured by X-ray diffraction. J Am Chem Soc 1952;74(6):1371-4.

[230] Norrish K. The swelling of montmorillonite. Discuss Faraday Soc 1954;18:120-34.

[231] Patel AD, Stamatakis E, Davis E. Shale hydration inhibition agent and method of use. US Patent 6247543, assigned to M-I LLC (Houston, TX); 2001. URL: http://www.freepatentsonline.com/6247543.html.

[232] Klein HP, Godinich CE. Drilling fluids. US Patent 7012043, assigned to Huntsman Petrochemical Corporation (The Woodlands, TX); 2006. URL: http://www.freepatentsonline.com/7012043.html.

[233] Tshibangu JP, Sarda JP, Audibert-Hayet A. A study of the mechanical and physicochemical interactions between the clay materials and the drilling fluids: Application to the boom clay (Belgium) (etude des interactions mecaniques et physicochimiques entre les argiles et les fluides de forage: Application a l'argile de boom (Belgique)). Rev Inst Franc Pet 1996;51(4):497-526.

[234] Suratman I. A study of the laws of variation (kinetics) and the stabilization of swelling of clay (contribution a l'etude de la cinetique et de la stabilisation du gonflement des argiles). Ph.D. thesis; Malaysia; 1985.

[235] Chen M, Chen Z, Huang R. Hydration stress on wellbore stability. In: Proceedings Volume. 35th US Rock Mech Symp. (Reno, NV, 6/5-7/95). ISBN 90-5410-552-6; 1995, p. 885-888.

[236] Tan CP, Richards BG, Rahman SS, Andika R. Effects of swelling and hydrational stress in shales on wellbore stability. In: Proceedings Volume. SPE Asia Pacific Oil & Gas Conf. (Kuala Lumpur, Malaysia, 4/14-16/97); 1997, p. 345-9.

[237] Clapper DK, Watson SK. Shale stabilising drilling fluid employing saccharide derivatives. EP Patent 0702073 assigned to Baker Hughes Incorporated; 1996. URL: https://www.google.at/patents/EP0702073A1?cl=en.

[238] Reid PI, Craster B, Crawshaw JP, Balson TG. Drilling fluid. US Patent 6544933, assigned to Schlumberger Technology Corporation (Sugar Land, TX); 2003. URL: http://www.freepatentsonline.com/6544933.html.

[239] McKenzie N, Ewanek J. Method and composition for cleaning and inhibiting solid, bitumin tar, and viscous fluid accretion in and on well equipment. US Patent 6564869, assigned to M-I, L.L.C. (Houston, TX); 2003. URL: http://www.freepatentsonline.com/6564869.html.

[240] Duncan Jr WM. Drilling fluid and drilling fluid additive. US Patent 5587354, assigned to Integrity Industries, Inc. (Kingville, TX); 1996. URL: http://www.freepatentsonline.com/5587354.html.

[241] Duncan Jr WM. Low toxicity terpene drilling fluid and drilling fluid additive. US Patent 5547925, assigned to Integrity Industries, Inc. (Kingsville, TX); 1996. URL: http://www.freepatentsonline.com/5547925.html.

[242] Chatterji J, Dealy ST, Crook RJ. Methods of removing water based drilling fluids and compositions. US Patent 6762155, assigned to Halliburton Energy Services, Inc. (Duncan, OK); 2004. URL: http://www.freepatentsonline.com/6762155.html.

[243] Deville J, Fritz B, Jarrett M. Development of water based drilling fluids customized for shale reservoirs. SPE Drill Complet 2011;26(4). doi:10.2118/140868-PA.

[244] Shenoy S, Gilmore T, Twynam A, Patel A, Mason S, Kubala G, et al. Guidelines for shale inhibition during openhole gravel packing with water based fluids. SPE Drill Complet 2008;23(2). doi:10.2118/103156-PA.

[245] Lee LJJ, Patel AD. Water based drilling fluids for reduction of water adsorption and hydration of argillaceous rocks. US Patent 5635458, assigned to M-I Drilling Fluids, L.L.C. (Houston, TX); 1997. URL: http://www.freepatentsonline.com/5635458.html.

[246] Ballard T, Beare S, Lawless T. Mechanisms of shale inhibition with water based muds. In: Proceedings Volume. IBC Tech. Serv. Ltd Prev. Oil Discharge from Drilling Oper. – The Options Conf. (Aberdeen, Scot, 6/23-24/93); 1993.

[247] Audibert A, Lecourtier J, Bailey L, Maitland G. Method for inhibiting reactive argillaceous formations and use thereof in a drilling fluid. WO Patent 1993015164 assigned to Inst Francais Du Petrole, Schlumberger Services Petrol, Schlumberger Technology Corp; 1993. URL: https://www.google.at/patents/WO1993015164A1?cl=en.

[248] Audibert A, Lecourtier J, Bailey L, Maitland G. Use of polymers having hydrophilic and hydrophobic segments for inhibiting the swelling of reactive argillaceous formations. EP Patent 0578806 assigned to Services Petroliers Schlumberger, Institut Francais Du Petrole; 1996. URL: https://www.google.at/patents/EP0578806B1?cl=en.

[249] Audibert A, Lecourtier J, Bailey L, Maitland G. Method for inhibiting reactive argillaceous formations and use thereof in a drilling fluid. US Patent 5677266, assigned to Inst. Francais Du Petrole; 1997.

[250] Jamaluddin AKM, Nazarko TW. Process for increasing near-wellbore permeability of porous formations. US Patent 5361845, assigned to Noranda, Inc. (Toronto, CA); 1994. URL: http://www.freepatentsonline.com/5361845.html.

[251] Reed MG. Permeability of fines-containing earthen formations by removing liquid water. CA Patent 2046792 assigned to Chevron Research And Technology Company, Marion G. Reed; 1993. URL: https://www.google.at/patents/CA2046792A1?cl=en.

[252] Kippie DP, Gatlin LW. Shale inhibition additive for oil/gas down hole fluids and methods for making and using same. US Patent 7566686, assigned to Clearwater International, LLC (Houston, TX); 2009. URL: http://www.freepatentsonline.com/7566686.html.

[253] Zeisel SH, da Costa KA. Choline: An essential nutrient for public health. Nutr Rev 2009;67(11):615-23. doi:10.1111/j.1753-4887.2009.00246.x.

[254] Eoff LS, Reddy BR, Wilson JM. Compositions for and methods of stabilizing subterranean formations containing clays. US Patent 7091159, assigned to Halliburton Energy Services, Inc. (Duncan, OK); 2006. URL: http://www.freepatentsonline.com/7091159.html.

[255] Elkatatny S, Mahmoud M, Nasr-El-Din H. Characterization of filter cake generated by water based drilling fluids using CT scan. SPE Drill Complet 2012;27(2). doi:10.2118/144098-PA.

[256] Davies SN, Meeten GH, Way PW. Water based drilling fluid additive and methods of using fluids containing additives. US Patent 5652200, assigned to Schlumberger Technology Corporation (Houston, TX); 1997. URL: http://www.freepatentsonline.com/5652200.html.

[257] Melear S, Guidroz Jr JA, Schlegel G, Micho W. Non-poluting anti-stick water-base drilling fluid modifier and method of use. US Patent 5120708, assigned to Baker Hughes Incorporated (Houston, TX); 1992. URL: http://www.freepatentsonline.com/5120708.html.

[258] Lummus JL, Scott Jr PP. Drilling mud system. US Patent 3223622, assigned to Pan American Petroleum Corp.; 1962. URL: http://www.freepatentsonline.com/3223622.html.

[259] Walker CO. Method for releasing stuck drill pipe. US Patent 4466486, assigned to Texaco Inc. (White Plains, NY); 1984. URL: http://www.freepatentsonline.com/4466486.html.

CHAPTER III

Fracturing Fluids

Fracture stimulation may be used in both vertical and horizontal wells. Fracturing horizontal wells may be undertaken in several situations, including situations where the formation has [1]:

1. restricted flow caused by low vertical permeability, the presence of shale streaks or formation damage;
2. low productivity due to low formation permeability;
3. natural fractures in a direction different from that of induced fractures, thus induced fractures have a high chance of intercepting the natural fractures; or
4. low stress contrast between the pay zone and the surrounding layers.

In the fourth case, a large fracturing treatment of a vertical well likely would not be an acceptable option since the fracture would grow in height as well as length. Drilling a horizontal well and creating either several transverse or longitudinal fractures may be preferable as they may allow rapid depletion of the reservoir through one or more fractures [1].

1. COMPARISON OF STIMULATION TECHNIQUES

In addition to hydraulic fracturing, there are other stimulation techniques such as acid fracturing or matrix stimulation. Hydraulic fracturing finds use not only in the stimulation of oil and gas reservoirs but also in coal seams to stimulate the flow of methane.

Fracturing fluids are often classified into water-based fluids, oil-based fluids, alcohol-based fluids, emulsion fluids, and foam-based fluids. Several reviews have been given in the literature dealing with the basic principles of hydraulic fracturing and the guidelines to select a particular formulation for a specific job [2–4].

Polymer hydration, crosslinking, and degradation are the key processes influencing their application. Technologic improvements over the years have focused primarily on improved rheologic performance, thermal stability, and cleanup of crosslinked gels.

Water-Based Chemicals and Technology for Drilling, Completion, and Workover Fluids
http://dx.doi.org/10.1016/B978-0-12-802505-5.00003-2

Copyright © 2015 Elsevier Inc.
All rights reserved.

2. SPECIAL TYPES

2.1 Viscoelastic Formulations

Viscoelastic surfactant (VES) fluids have advantages over conventional polymer formulations. These include [5]:

- higher permeability in the oil-bearing zone;
- lower formation or subterranean damage;
- higher viscosifier recovery after fracturing;
- no need for enzymes or oxidizers to break down viscosity; and
- easier hydration and faster buildup to optimum viscosity.

Disadvantages and drawbacks of VES fluids are the high costs of the surfactants, their low tolerance to salts, and stability against high temperatures as found in deep well applications. However, there are recent formulations that overcome these difficulties, at least to some extent.

The components of a viscoelastic fluids are a zwitterionic surfactant, erucyl amidopropyl betaine, an anionic polymer, or N-erucyl-N,N-bis(2-hydroxyethyl)-N-methyl ammonium chloride, poly(napthalene sulfonate), and cationic surfactants, methyl poly(oxyethylene) octadecaneammonium chloride and poly(oxyethylene) cocoalkylamines [5, 6]. The corresponding fluids exhibit a good viscosity performance.

Typical VESs are N-erucyl-N,N-bis(2-hydroxyethyl)-N-methyl ammonium chloride and potassium oleate, solutions of which form gels when mixed with corresponding activators such as sodium salicylate and potassium chloride [7].

The cationic surfactant should be soluble in both organic and inorganic solvents. Typically, a high solubility of the cationic surfactant in hydrocarbon solvents is promoted by multiple long chain alkyl groups attached to the active surfactant unit [7]. Examples are hexadecyltributylphosphonium and trioctylmethylammonium ions. In contrast, cationic surfactants for viscoelastic solutions have rather a single long linear hydrocarbon moiety attached to the surfactant group. Obviously, there is a conflict between the structural requirements for achieving solubility in hydrocarbons and for the formation of viscoelastic solutions.

As a compromise, surfactant compounds that are suitable for reversibly thickening water-based wellbore fluids and also soluble in both organic and aqueous fluids have been designed.

A non-ionic surfactant gelling agent is tallow amido propylamine oxide [8]. Non-ionic fluids are inherently less damaging to the producing formations than cationic fluid types and are more efficacious than anionic gelling agents.

The synthesis of branched oleate has been described. 2-Methyl oleic acid methyl ester also addressed as 2-methyl oleate can be prepared from methyl oleate and methyl iodide in presence of a pyrimidine-based catalyst [7]. The methyl ester is then hydrolyzed to obtain 2-methyl oleic acid.

It is sometimes believed that the contact of a VES-gelled fluid instantaneously reduces the viscosity of the gel. However, it has been discovered that mineral oil can be used as an internal breaker for VES-gelled fluid systems [9].

The rate of breaking the viscosity at a certain temperature is influenced by the type and amount of salts present. In the case of low-molecular weight mineral oils, it is important to add them after the VES component is added to the aqueous fluid.

By using combinations of internal breakers, both the initial and final break of the VES fluid may be customized. In addition to mineral oil, for instance, fatty acid compounds or bacteria may be used [9, 10].

2.2 Miscellaneous Polymers

A copolymer of $2''$-ethylhexyl acrylate and AA is not soluble either in water or in hydrocarbons. The ester units are hydrophobic and the acid units are hydrophilic. An aqueous suspension with a particle size smaller than 10 μm can be useful in preparing aqueous hydraulic fracturing fluids [11]. $2''$-Ethylhexyl acrylate is shown in Figure 23.

A water-soluble polymer from N-vinyl lactam monomers or vinyl-containing sulfonate monomers reduces the water loss and enhances other properties of well treating fluids in high-temperature subterranean environments [12]. Lignites, tannins, and asphaltic materials are added as dispersants. Vinyl monomers for modification are shown in Figure 24.

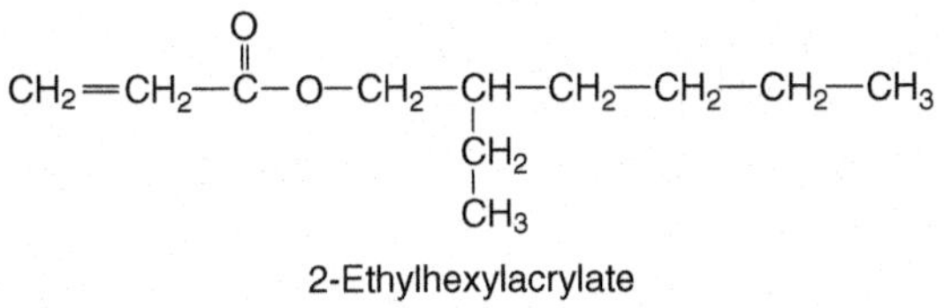

2-Ethylhexylacrylate

Figure 23 $2''$-Ethylhexyl acrylate.

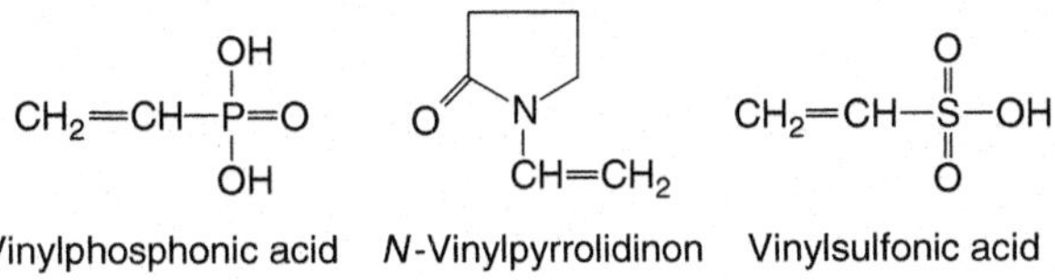

Vinylphosphonic acid   N-Vinylpyrrolidinon   Vinylsulfonic acid

Figure 24 Monomers for synthetic thickeners.

2.2.1 Degradable Polymers

Degradable thermoplastic lactide polymers are used for fracturing fluids. Hydrolysis is the primary mechanism for degradation of the lactide polymer [13].

2.2.2 Biodegradable Formulations

Biodegradable fluid formulations have been suggested. These are formulations of a poly(saccharide) in a concentration insufficient to permit a contaminating bacterial proliferation, namely a high-viscosity carboxy methyl cellulose (CMC) sensitive to bacterial enzymes produced by the degradation of the poly(saccharide) [14].

On the other hand, the biodegradability of mud additives is a problem [15]. The biodegradability of seven kinds of mud additives was studied by determining the content of dissolved oxygen in water, a simple biochemical oxygen demand testing method. The biodegradability is high for starch but is lower for polymers of allyl monomers and additives containing an aromatic group.

2.3 Concentrates

Historically, fracture stimulation treatments have been performed by using conventional batch mix techniques. This involves premixing chemicals into tanks and circulating the fluids until a desired gelled fluid rheology is obtained. This method is time-consuming and burdens the oil company with disposal of the fluid if the treatment ends prematurely.

Environmental concerns, such as when spillage occurs or disposal is involved, can be avoided if the fluid is capable of being gelled as needed. Thus a newer technology involves a gelling-as-needed technology with water, methanol, and oil [16]. This procedure eliminates batch mixing and minimizes handling of chemicals and base fluid. The customer is charged only for products used, and environmental concerns regarding disposal are virtually eliminated. Computerized chemical addition and monitoring, combined with on-site procedures, ensure quality control throughout a treatment. Fluid rheologies can be accurately varied during the treatment by varying polymer loading.

The use of a diesel-based concentrate with hydroxypropyl guar gum has been evolved from the batch-mixed dry powder types [17]. The application of such a concentrate reduces system requirements. Companies can benefit from the convenience of the reduced logistic burden that comes from using the diesel hydroxypropyl guar gum concentrate.

Table 34 Components of a slurry concentrate

Component	Example
Hydrophobic solvent base	Diesel
Suspension agent	Organophilic clay
Surfactant	Ethoxylated nonyl phenol
Hydratable polymer	Hydroxypropyl guar gum

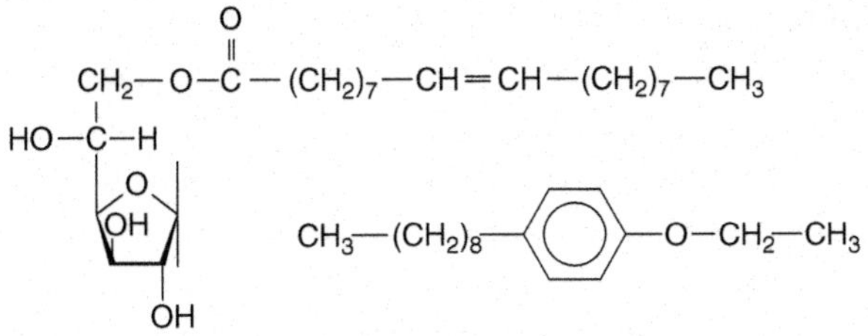

Sorbitan monooleate   Ethoxylated nonylphenol

Figure 25 Surfactants in a slurry concentrate.

A fracturing fluid slurry concentrate has been proposed [18] that consists of the components shown in Table 34. Such a polymer slurry concentrate will readily disperse and hydrate when admixed with water at the proper pH, thus producing a high-viscosity aqueous fracturing fluid. The fracturing fluid slurry concentrate is useful in producing large volumes of high-viscosity treating fluids at the well site on a continuous basis. Surfactants in a slurry concentrate are shown in Figure 25.

Fluidized aqueous suspensions of 15% or more of hydroxyethyl cellulose (HEC), hydrophobically modified cellulose ether, hydrophobically modified HEC, methyl cellulose, hydroxypropylmethyl cellulose, and poly(ethylene oxide) are prepared by adding the polymer to a concentrated sodium formate solution containing xanthan gum as a stabilizer [19].

The xanthan gum is dissolved in water before sodium formate is added. Then the polymer is added to the solution to form a fluid suspension of the polymers. The polymer suspension can serve as an aqueous concentrate for further use.

2.4 Foamed Fracturing Fluids

Increasing the conductivity of a formation is known as *fracturing* the formation. Conventionally, in a fracturing process, a fluid is injected into the formation at sufficient pressure and rate to fracture the formation.

The fracturing fluid has of sufficient viscosity to carry a proppant into the formation and to obtain fracturing of the formation at high pressure, shear, and temperature. The proppant consists of such materials as sand, glass beads, sintered bauxite, fine gravel, and the like. The proppant is carried into the fracture and helps keep the formed fractures open after fluid recovery. The newly formed fracture is held open by proppant which provides increased routes or channels through which hydrocarbons can flow increasing production.

It has been found that foams could be used in place of the known slurries, emulsions, and gels [20]. The foams are composed from nitrogen, carbon dioxide, and water with a suitable surfactant. The foam composition with the slurries, gels, and emulsions, and proppant is pumped as usually into the formation at such a pressure that it causes the fracture of the formation. However, unlike the conventional fracturing systems, the use of foams increases the well cleanup because the foam is easily removed from the well, when the pressure is reduced.

In making such a foam, a fracturing gel is first made from water or brine, a gelling agent, that is, a polymer and a suitable surfactant. The water or brine may contain up to about 20% alcohol, for example, methanol. In addition, the fracturing gel may be crosslinked by the use crosslinking agents.

After the gel is made, a foaming agent is added and the gel with the foaming agent is added to an energizing phase such as carbon dioxide or nitrogen or a combination of carbon dioxide and nitrogen to create the foam.

When a fracturing gel is based on a slurried polymer system, this can be crosslinked. Such a system has the advantage of being able to supply the gelling agent polymer in liquid form to the aqueous fluid as the aqueous fluid is pumped into the formation. If the fracturing job is discontinued, then the supply of polymer is simply stopped, thus, the user does not have to face disposal problems of unused gel.

However, in some cases, the slurried polymer system is not compatible with fracturing gels that are using alkylbetaines as the foamer [21]. Slurried polymer components that do not have this drawback are high–molecular weight solvatable poly(saccharide)s. A preferred polysaccharide is sodium carboxymethylhydroxypropyl guar.

Environmentally safe foamed fracturing fluids have been described [22, 23]. These fracturing fluids basically contain water, a gelling agent, sufficient gas to form a foam, and an additive for foaming and stabilizing the gelled water comprised of hydrolyzed keratin.

The gelling agent is a cellulose derivative selected from HEC, carboxymethylhydroxyethyl cellulose, or HEC grafted with vinylphosphonic acid. A hydrolyzed keratin is used for foaming.

This can be manufactured by the hydrolysis of hoof and horn meal. Hoof and horn meal is heated with lime in an autoclave to produce a hydrolyzed protein. The protein is commercially available as a free flowing powder that contains about 85% protein.

2.5 Low Friction Fracturing Fluid

An organic boron zirconium crosslinking agent was synthesized and formulated as a low friction type fracturing fluid [24]. The performance of this fracturing fluid was evaluated in the laboratory. The composition will perform well at high temperatures. A viscosity of 100 mP or more after 90 min of shearing with 170 s^{-1} at $180\,°C$ is maintained.

Furthermore, the composition exhibits a delayed crosslinking; 120 s is needed to complete the crosslinking reaction. Therefore, the tube friction is reduced to a large extent.

Also, low friction pressures were observed, as the drag reduction percentage was in the range of 35-70%. Only a minor damage was caused by the fluid to the formation core samples. The average core permeability damage was 19.6% [24].

2.6 Slickwater Fracturing

One of the challenges in slickwater fracturing of tight sand gas reservoirs is post-treatment fluid recovery [25]. More than 60% of the injected fluid remains in the critical near wellbore area and has a significant negative impact on the relative permeability to gas and well productivity. The trapped water could be due to capillary forces around the vicinity of the fractured formation. For strongly water-wet tight gas reservoirs, capillary forces promote the retention of injected fluids in pore spaces.

Commonly available surfactants are added to slickwater to reduce surface tension between the treating fluids and gas. The problem with surfactants is that upon exposure to the formation, they adsorb on the surface of the rock.

The addition of a microemulsion to the fracturing fluid can result in lowering the pressure needed to displace injected fluids or condensate from low-permeability core samples. This alteration of the fracturing fluid effectively lowers the capillary forces in low-permeability reservoirs. This will result in removal of water and condensate blocks, the mitigation of phase trapping, and therefore an increase in permeability to gas.

The effectiveness of microemulsions in the improvement of fracturing fluid recovery have been examined. An environmental friendly microemulsion formulated with a blend of an anionic surfactant, non-ionic surfactant, short chain alcohol, and water showed very good results in lowering interfacial tension between water and oil, in comparison to competitive technologies.

The performance of this microemulsion was excellent in high salinity fluid as well as low salinity fluid [25].

2.7 Environmental Friendly Fracturing

A non-toxic, environmental friendly, green flowback aid has been developed that reduces water blockage when injected into a fractured reservoir. This environmentally safe composition is derived from renewable resources and contains [26]:

- water-soluble esters of a low-molecular weight alcohol and a low-molecular weight organic acid;
- oil-soluble esters of a low-molecular weight alcohol and a high-molecular weight fatty acid;
- water-soluble or dispersible non-ionic surfactants derived from vegetable or animal sources;
- anionic or amphoteric surfactants derived from animal- or vegetable-based sources; and
- water.

Examples of esters and surfactants are given in Table 35.

3. SPECIAL ADDITIVES FOR FRACTURING FLUIDS

A general review of commercially available additives for fracturing fluids is given in the literature [27]. Possible components in a fracturing fluid are listed in Table 36.

In particular, Table 36 reflects the complexity of a fracturing fluid formulation. Some additives may not be used together reasonably such as oil-gelling additives in a water-based system. More than 90% of the

Table 35 Esters and surfactants [26]

Water soluble	Organo soluble
Ethyl acetate	Methyl lactate
Propyl acetate	Propyl lactate
Butyl acetate	Methyl succinate
Ethyl lactate	Ethyl succinate
Organo soluble	**Organo soluble**
Ethyl laurate	Methyl soyate
Methyl laurate	Ethyl soyate
Propyl laurate	Propyl soyate
Methyl oleate	Methyl cocoate
Ethyl oleate	Ethyl cocoate
Propyl oleate	Propyl cocoate
Methyl racinoleate	Methyl erucate
Ethyl racinoleate	Ethyl erucate
Propyl racinoleate	Propyl erucate
Non-ionic surfactant	**Non-ionic surfactant**
Oxirane	Palmityl alcohol
Methyl oxirane	Oleyl alcohol
Lauryl alcohol	Erucyl alcohol
Myristyl alcohol	Polyglucosides

fluids are based on water. Aqueous fluids are economical and can provide control of a broad range of physical properties as a result of additives developed over the years. Additives for fracturing fluids serve two purposes [28]:

1. to enhance fracture creation and proppant-carrying capability and
2. to minimize formation damage.

Additives that assist the creation of a fracture include viscosifiers such as polymers and crosslinking agents, temperature stabilizers, pH control agents, and fluid loss control materials. Formation damage is reduced by such additives as gel breakers, biocides, surfactants, clay stabilizers, and gases. Table 37 summarizes the various types of fluids and techniques used in hydraulic fracturing.

3.1 Thickeners and Gelling Agents

A gelling agent is also addressed as a viscosifying agent. A gelling agent refers to a material capable of forming the fracturing fluid into a gel, thereby increasing its viscosity [31].

Table 36 Components in fracturing fluids

Component/category	Function/remark
Water-based polymers	Thickener, to transport proppant, reduces leak-off in formation
Friction reducers	Reduce drag in tubing
Fluid loss additives	Form filter cake, reduce leak-off in formation if thickener is not sufficient
Breakers	Degrade thickener after job or disable crosslinking agent (wide variety of different chemical mechanisms)
Emulsifiers	For diesel premixed gels
Clay stabilizers	For clay-bearing formations
Surfactants	Prevent water-wetting of formation
Nonemulsifiers	
pH control additives	Increase the stability of fluid (e.g., for elevated temperature applications)
Crosslinking agents	Increase the viscosity of the thickener
Foamers	For foam-based fracturing fluids
Gel stabilizers	Keep gels active longer
Defoamers	
Oil-gelling additives	Same as crosslinking agents for oil-based fracturing fluids
Biocides	Prevent microbial degradation
Water-based gel systems	Common
Crosslinked gel systems	Increase viscosity
Alcohol-water systems	
Oil-based systems	Used in water-sensitive formation
Polymer plugs	Used also for other operations
Gel concentrates	Premixed gel on diesel base
Resin-coated proppants	Proppant material
Ceramics	Proppant material

Table 37 Various types of hydraulic fracturing fluids

Type	Remarks
Water-based fluids	Predominant
Oil-based fluids	Water-sensitive; increased fire hazard
Alcohol-based fluids	Rarely used
Emulsion fluids	High pressure, low temperature
Foam-based fluids	Low pressure, low temperature
Noncomplex gelled water fracture	Simple technology
Nitrogen-foam fracture	Rapid cleanup
Complexed gelled water fracture	Often the best solution
Premixed gel concentrates	Improve process logistics
In situ precipitation technique	Reduces scale-forming ingredients [29, 30]

Suitable gelling agents include guar gum, xanthan gum, welan gum, locust bean gum, gum ghatti, gum karaya, tamarind gum, and tragacanth gum. A suitable depolymerized gum is a depolymerized guar gum. The guar gum can be functionalized of modified to result in hydroxyethyl guar, hydroxypropyl guar, and carboxymethyl guar. Examples of water-soluble cellulose ethers include methyl cellulose, CMC, HEC, and hydroxyethyl carboxymethyl cellulose [31].

Artificial polymers are as copolymers from AAm, methacrylamide, AA, or maleic anhydride (MA). Other types are copolymers from acrylo amido propane sulfonate (AMPS) derivates and N-vinylpyridine (NVP) [31].

Naturally occurring poly(saccharide)s and their derivatives form the predominant group of water-soluble species generally used as thickeners to impart viscosity to treating fluids [32]. Other synthetic polymers and biopolymers have found ancillary applications. Polymers increase the viscosity of the fracturing fluid in comparatively small amounts. The increase in fluid viscosity of hydraulic fracturing fluids serves for improved proppant placement and fluid loss control. Table 38 summarizes polymers suitable for fracturing fluids.

Guar is shown in Figure 26. In hydroxypropyl guar, some of the hydroxyl groups are etherified with oxopropyl units. Compositions for gelling a hydrocarbon fracturing fluid are basically different from those for aqueous fluids. A possible formulation consists of a gelling agent, a phosphate ester, a crosslinking agent, a multivalent metal ion, a catalyst, and a fatty quaternized amine [37].

Table 38 Summary of thickeners suitable for fracturing fluids

Thickener	References
Hydroxypropyl guar[a]	
Galactomannans[b]	[33]
HEC-modified vinylphosphonic acid	[34]
Carboxymethyl cellulose	
Polymer from N-vinyl lactam monomers, vinylsulfonates[c]	[12]
Reticulated bacterial cellulose[d]	[35]
Bacterial xanthan[e]	[36]

[a]General purpose eightfold power of thickening in comparison to starch.
[b]Increased temperature stability, used with boron-based crosslinking agents.
[c]High-temperature stability.
[d]Superior fluid performance.
[e]Imparts high viscosity.

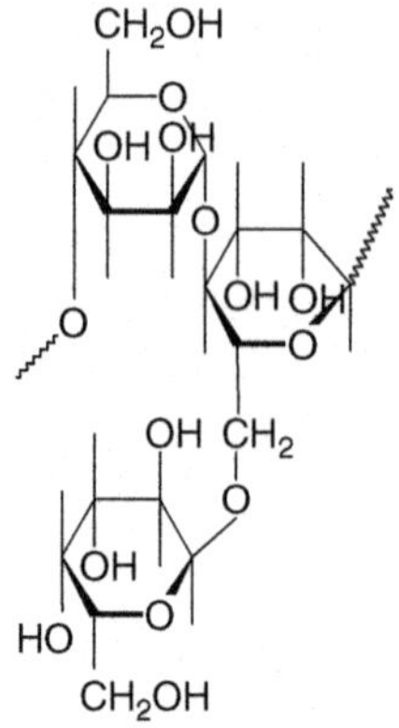

Figure 26 Structural unit of guar.

3.1.1 Guar

Fracturing fluids have traditionally been viscosified with guar and guar derivatives. Guar-based swelling agents normally used in the form of more or less fine powders have extremely effective thickening properties when dissolved in aqueous phases. There are reliable methods for controlling the viscosity of correspondingly thickened aqueous treatment solutions using both products based on guar or its derivatives and, suitable crosslinking agents, which can additionally control the viscosity in the aqueous phase through non-ionic, anionic, or cationic mechanisms [38].

Actually, guar is a branched poly(saccharide) from the guar plant *Cyamopsis tetragonolobus*, originally from India, now also found in the southern United States, with a molar mass of approximately 220 kDa. It consists of mannose in the main chain and galactose in the side chain. The ratio of mannose to galactose is 2:1.

Poly(saccharide)s with a mannose backbone and side chains unlike mannose are referred to as *heteromannans* according to the nomenclature of poly(saccharide)s, in particular as *galactomannans*. Derivatives of guar therefore are sometimes called *galactomannans*.

Guar-based gelling agents, typically hydroxypropyl guar, are widely used to viscosify fracturing fluids because of their desirable rheologic properties, economics, and ease in hydration. Nonacetylated xanthan is a variant of xanthan gum, which develops a synergistic interaction with guar that exhibits a superior viscosity and particle transport at lower polymer concentrations.

Static leak-off experiments with borate-crosslinked and zirconate-crosslinked hydroxypropyl guar fluids showed practically the same leak-off coefficients [39]. An investigation of the stress-sensitive properties showed that zirconate filter cakes have viscoelastic properties, but borate filter cakes are merely elastic. Noncrosslinked fluids show no filter cake-type behavior for a large range of core permeabilities, but rather a viscous flow dependent on porous medium characteristics.

The addition of glycols, such as ethylene glycol (EG), to aqueous fluids gelled with a guar gum compound can increase the viscosity of the fluid and stabilize the fluid brines. In particular, the gelled aqueous fluids are more stable at high temperatures such as in the range of from 27 to 177 °C (80–350 °F).

This finding allows the guar to be used at high temperatures. The formation damage is minimized after hydraulic fracturing operations as less of the guar polymer can be used, but the same viscosity is achieved by the addition of a glycol [40].

The crosslinking agent is selected from borates, titanates, or zirconates. The stability of the gel is improved by the addition of sodium thiosulfate. The development of the viscosity at 93 °C (200 °F) of brine fluids with 2.4 $kg\,m^{-3}$ guar and 5% KCl, with varying amounts of EG is shown in Figure 27.

By using Na ethylene diamine tetraacetic acid (EDTA) as gel breaker in these compositions with EG, the decay of viscosity with time can be still adjusted accordingly [40].

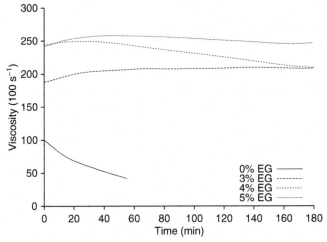

Figure 27 Viscosity of guar brines with varying amounts of ethylene glycol (EG) [40].

Anionic galactomannans, which are derived from guar gum, in which the hydroxyl groups are partially esterified with sulfonate groups that result from AMPS and 1-allyloxy-2-hydroxypropyl sulfonic acid [41], have been claimed to be suitable as thickeners. The composition is capable of producing enhanced viscosities, when used either alone or in combination with a cationic polymer and distributed in a solvent.

Polyhydroxy compounds can be modified by various reactions. The primary hydroxyl group in the C_6 position is particularly suitable for derivatization, although secondary hydroxyl groups of the polymer skeleton may also be derivatized [38].

Etherification, exemplified with dextrose as the model compound is shown in Figure 28. Vinyl compounds, specifically used for the modification of guar, are shown in Figure 29.

The addition of ester oils show a particularly pronounced stabilizing effect in a highly concentrated guar mixture. The interaction between the ester group on the one hand and the highly hydrophilic outer surface of the

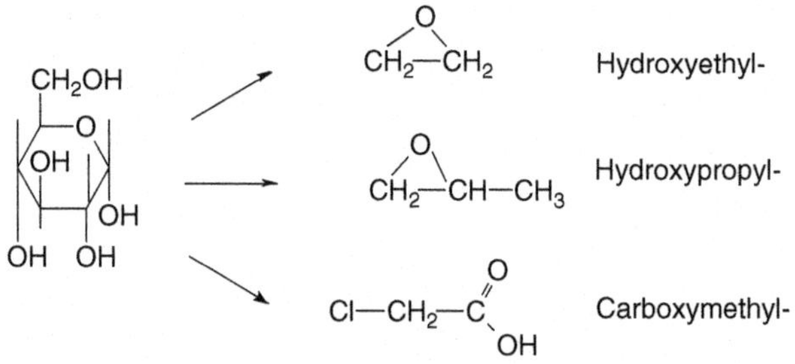

Figure 28 Modification of polyhydroxy compounds.

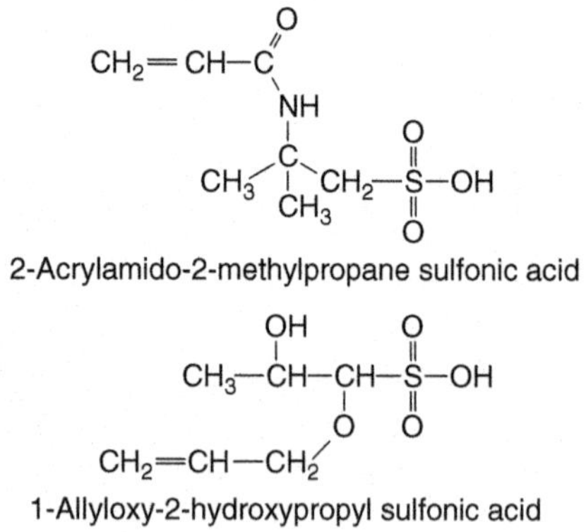

Figure 29 Vinyl modifiers for guar gum.

guar-based solid particle on the other hand would appear to be particularly intensive [38].

Borate-crosslinked galactomannan fracturing fluids have an increased temperature stability. The temperature stability of fracturing fluids containing galactomannan polymers is increased by adding a sparingly soluble borate with a slow solubility rate to the fracturing fluid. This provides a source of boron for solubilizing at elevated temperatures, thus enhancing the crosslinking of the galactomannan polymer. The polymer also improves the leak-off properties of the fracturing fluid.

3.1.2 Hydroxyethyl Cellulose

HEC can be chemically modified by the reaction with vinylphosphonic acid in the presence of the reaction product of hydrogen peroxide and a ferrous salt. The HEC forms a graft copolymer with the vinylphosphonic acid.

Amylose and cellulose are shown in Figure 30. Amylose consists of a water-soluble portion, a linear polymer of glucose, the amylose; and a water-insoluble portion, the amylopectin. The difference between amylose and cellulose is the way in which the glucose units are linked. In amylose, α-linkages are present, whereas in cellulose, β-linkages are present. Because of this difference, amylose is soluble in water and cellulose is not. Chemical modification allows cellulose to become water soluble.

Modified HEC has been proposed as a thickener for hydraulic fracturing fluids [34]. Polyvalent metal cations may be employed to crosslink the polymer molecules to further increase the viscosity of the aqueous fluid.

3.1.3 Biotechnological Products

Gellan Gum and Welan Gum

Gellan gum is the generic name of an extracellular poly(saccharide) produced by the bacterium *Pseudomonas elodea*. Gellan gum is a linear anionic

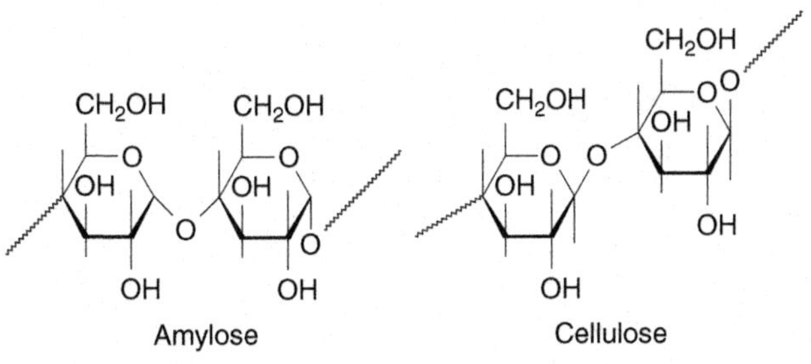

Figure 30 Amylose and cellulose.

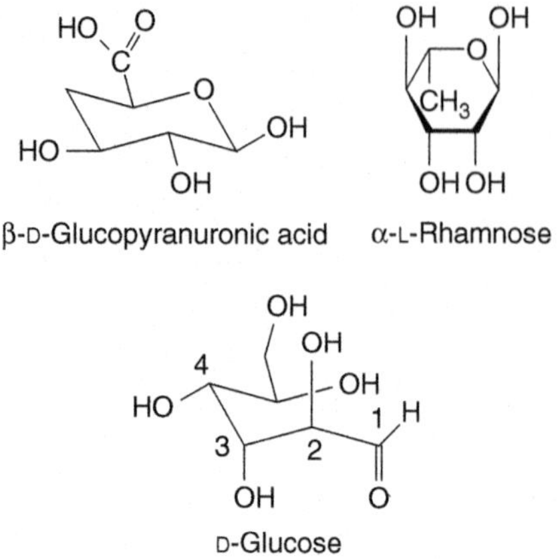

β-ᴅ-Glucopyranuronic acid   α-ʟ-Rhamnose

ᴅ-Glucose

Figure 31 Gellan gum monomers.

poly(saccharide) with a molecular mass of 500 kDa. It is composed of 1,3-β-ᴅ-glucose, 1,4-β-ᴅ-glucuronic acid, 1,4-β-ᴅ-glucose, and 1,4-α-ʟ-rhamnose. These compounds are shown in Figure 31.

Welan gum is produced by aerobic fermentation. The backbone of welan gum is identical to gellan gum, but it has a side chain that consists of ʟ-mannose or ʟ-rhamnose. It is used in fluid loss additives and is extremely compatible with calcium ions in alkaline solutions.

Reticulated Bacterial Cellulose

A cellulose with an intertwined reticulated structure, produced from bacteria, has unique properties and functionalities unlike other conventional celluloses. When added to aqueous systems, reticulated bacterial cellulose improves the fluid rheology and the particle suspension over a wide range of conditions [35]. Test results showed advantages in fluid performance and significant economic benefits by the addition of reticulated bacterial cellulose.

Xanthan Gum

Xanthan gum is produced by the bacterium *Xanthomonas campestris*. Commercial productions started in 1964. Xanthans are water-soluble poly(saccharide) polymers with the repeating units [42], as given in Table 39 and Figure 32.

The ᴅ-glucose moieties are linked in a β-(1,4) configuration. The inner ᴅ-mannose moieties are linked in an α-(1,3) configuration, generally to alternate glucose moieties. The ᴅ-glucuronic acid moieties are linked

Table 39 Variant xanthan gums

Number	Repeating units	Ratio
Pentamer	D-glucose:D-mannose:D-glucuronic acid	2:2:1
Tetramer	D-glucose:D-mannose:D-glucuronic acid	2:1:1

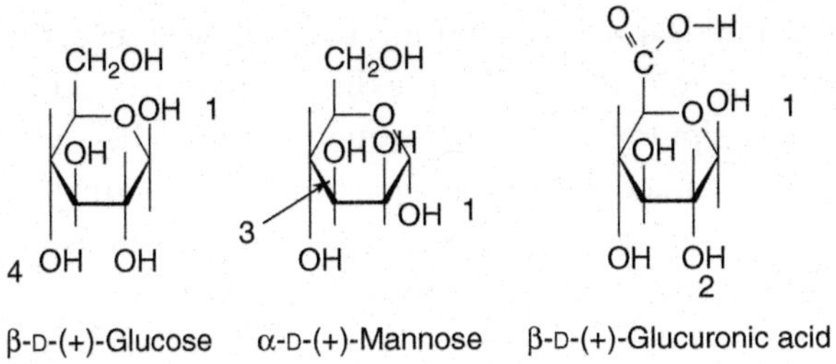

β-D-(+)-Glucose   α-D-(+)-Mannose   β-D-(+)-Glucuronic acid

Figure 32 Carbohydrates and derivates.

in a β-(1,2) configuration to the inner mannose moieties. The outer mannose moieties are linked to the glucuronic acid moieties in a β-(1,4) configuration.

Most of the xanthan gum used in oil field applications is in the form of a fermentation broth containing 8-15% polymer. The viscosity is less dependent on the temperature in comparison with other poly(saccharide)s.

3.2 Friction Reducers

Low pumping friction pressures are achieved by delaying the crosslinking, but there are also specific additives available to reduce the drag in the tubings. The first application of drag reducers was using guar in oil well fracturing, now a routine practice.

Relatively small quantities of a bacterial cellulose ($0.60-1.8$ g l^{-1}) in hydraulic fracturing fluids enhance their rheologic properties [43]. The suspension of the proppant is enhanced and friction loss through well casings is reduced.

3.3 Fluid Loss Additives

Fluid loss additives are used widely as additives for drilling and fracturing fluids. When fracturing zones have high permeability, concern exists about damage to the matrix from deeply penetrating fluid leak-off along the fracture or caused by the materials in the fluid that minimize the amount of leak-off.

Several fracturing treatments in high-permeability formations, which are characterized by short lengths, and often by disproportionate widths, exhibit positive post-treatment skin effects. This is the result of fracture face-damage [44]. If the invasion of the fracturing fluid is minimized, the degree of damage is of secondary importance.

Thus, if the fluid leak-off penetration is small, even severe permeability impairments can be tolerated without exhibiting positive skin effects. The first priority in designing fracture treatments should be maximizing the conductivity of the fracture. In high-permeability fracturing, the use of high concentrations of polymer-crosslinked fracturing fluids with fluid loss additives and breakers is recommended.

Materials used to minimize leak-off also have the potential to damage the conductivity of the proppant pack. High shear rates at the tip of the fracture may prevent the formation of external filter cakes, increasing the magnitude of spurt losses in highly permeable formations. Therefore, particularly for fracturing tasks, non-damaging additives are needed. Enzymatically degradable fluid loss additives are available. Table 40 summarizes some fluid loss additives suitable for hydraulic fracturing fluids.

Fluid loss additives that are containing microgels have been described [51]. A polymeric microgel refers to a gelled particle comprising a crosslinked polymer, that is, a water-soluble and water-swellable network.

It is beneficial to include bridging agents in the drilling fluid that will become incorporated into the filter cake. Without the bridging agents, the polymeric microgels may form an effective filter cake, but that filter cake may be more difficult to remove.

Table 40 Fluid loss additives for hydraulic fracturing fluids

Chemical	References
Calcium carbonate and lignosulfonate[a]	[45, 46]
Natural starch	[47–49]
Carboxymethyl starch	[47–49]
Hydroxypropyl starch[b]	[47–49]
HEC with crosslinked guar gums[c]	[50]
Granular starch and particulate mica	[50]

[a]Welan or xanthan gum polymer can be added to keep the calcium carbonate and lignosulfonate in suspension.
[b]Synergistic effect, see text.
[c]500 mD permeability.

Polymeric microgels are a polymerization product from a suitable monomer and a crosslinking agent. A special method forming the microgels is the dispersion polymerization in a continuous medium.

Polymeric microgel particles can be prepared from the copolymerization of acrylamide and diethylene glycol dimethacrylate in ethanol in the presence of poly(vinylpyrrolidone). Instead of diethylene glycol dimethacrylate, N,N'-ethylene-bis-acrylamide can be used as crosslinking agent. As initiator, 2,2'-azobisisobutyronitrile is added. Several other methods for the preparation of microgels have been described in detail [51].

Crosslinked polymer microgel particles are formed, which are believed to be insoluble or at most swellable in the continuous medium. It is believed that the crosslinking agent may act as an initiator for chain branching, which in turn may react with one another to form a polymeric microgel [51].

A water-based fluid loss additive containing an amphiphilic dispersant has been disclosed [52]. The fluid loss additive is a water-swellable AMPS random copolymer, and the dispersant is a hydrotrope of sodium cumene sulfonate.

The ability of a hydrotrope to increase the water solubility of some compounds is known as hydrotropy. Hydrotropes contain both a hydrophilic and a hydrophobic functional group. The hydrophobic part of the molecule is a benzene substituted apolar segment. For example, the hydrophobic part can be N-butyl benzene, dodecyl benzene, xylene, and cumene. The hydrophilic, polar segment is an anionic sulfonate group accompanied by a counter ion, for example, ammonium, calcium, potassium, or sodium.

Hydrotropes are produced by the sulfonation of an aromatic hydrocarbon solvent; for example, toluene, xylene, or cumene. The resulting aromatic sulfonic acid is neutralized using an appropriate base, for example, sodium hydroxide to produce the sulfonate [52].

In general, the hydrophobic part of a hydrotrope is too small to cause a spontaneous self-aggregation. The self-aggregation reaction is similar to the micelle formation for a surfactant. However, some hydrotropes can self-aggregate. Self-aggregation of a hydrotrope is generally formed in a stepwise process, with the size of the aggregate gradually increasing.

Unlike the critical micelle concentration for a surfactant, a hydrotrope generally does not have a critical concentration at which self-aggregation occurs. Rather, the ability of a hydrotrope to self-aggregate appears to be more related to the chemical structure of the hydrotrope [52].

3.3.1 Degradation of Fluid Loss Additives

A fluid loss additive for fracturing fluids, which is a mixture of natural starch (corn starch) and chemically modified starches (carboxymethyl and hydroxypropyl derivatives) plus an enzyme, has been described [53, 54]. The enzyme degrades the α-linkage of starch but does not degrade the β-linkage of guar and modified guar gums when used as a thickener.

3.3.2 Permeable Fracturing Additive

A fracturing material has been described that comprises a viscous curable support material and fibers embedded in the support material [55]. In this way, a solid support matrix with permeable channels is formed by the fibers that are embedded in the cured support material.

The fibers may be synthetic fibers or natural fibers. Poly(tetrafluorethylene), nylon, polyester, acrylic, and polyolefin made fibers are examples for synthetic fibers. Synthetic fibers are man-made, whereas natural fibers are fibers which are found in nature. Poly(tetrafluorethylene) fibers have turned out to be particularly advantageous, since when being mixed with the curable support material such as cement. There the poly(tetrafluorethylene) fibers are capable to form a channel network allowing for a high degree of permeability.

Synthetic fibers may be created by forcing, for instance through extrusion, fiber forming materials through holes into the air, forming a thread. Natural fibers may come from plants or animals. An example for such natural fibers are cellulose fibers [55].

3.3.3 Granular Starch and Mica

A fluid loss additive has been described that consists of granular starch composition and fine particulate mica [50]. An application comprises a fracturing fluid containing this additive. A method of fracturing a subterranean formation penetrated by a borehole comprises injecting into the borehole and into contact with the formation, at a rate and pressure sufficient to fracture the formation, a fracturing fluid containing the additive in an amount sufficient to provide fluid loss control.

3.3.4 Depolymerized Starch

Partially depolymerized starch provides decreased fluid losses at much lower viscosities than the corresponding starch derivatives that have not been partially depolymerized [56].

3.3.5 Controlled Degradable Fluid Loss Additives

A fluid loss additive for a fracturing fluid comprises a mixture of natural and modified starches plus an enzyme [54]. The enzyme degrades the α-linkage of starch but does not degrade the β-linkage of guar and modified guar gums when used as a thickener.

Natural or modified starches are utilized in a preferred ratio of 3:7 to 7:3, with optimum at 1:1, and the mix is used in the dry form for application from the surface down the well. The preferred modified starches are the carboxymethyl and hydroxypropyl derivatives. Natural starches may be those of corn, potatoes, wheat, and soy, and the most preferred is corn starch.

Blends include two or more modified starches, as well as blends of natural and modified starches. The starches can be coated with a surfactant, such as sorbitan monooleate, ethoxylated butanol, or ethoxylated nonyl phenol, to facilitate the dispersion into the fracturing fluid. Modified starches or blends of modified and natural starches with a broad particulate size distribution have been found to maintain the injected fluid within the created fracture more effectively than natural starches [57]. The starches can be degraded by oxidation or by bacterial attack.

A fluid loss additive is described [57] that helps to achieve a desired fracture geometry by lowering the spurt loss and leak-off rate of the fracturing fluid into the surrounding formation by rapidly forming a filter cake with low permeability. The fluid loss additive is readily degraded after the completion of the fracturing process. The additive has a broad particulate size distribution that is ideal for use in effectively treating a wide range of formation porosities and is easily dispersed in the fracturing fluid.

The fluid loss additive comprises a blend of modified starches or blends of one or more modified starches and one or more natural starches. The additive is subject to controlled degradation to soluble products by a naturally proceeding oxidation reaction or by bacterial attack by bacteria naturally present in the formation. The oxidation may be accelerated by adding oxidizing agents such as persulfates and peroxides.

3.4 pH Control Additives

Buffers are necessary to adjust and maintain the pH. Buffering agents can be salts of a weak acid and a weak base. Examples are carbonates, bicarbonates, and hydrogen phosphates [58]. Weak acids such as formic acid, fumaric acid, and sulfamic acid also are recommended. Common aqueous buffer ingredients are shown in Table 41 and in Figures 33 and 34.

Table 41 Common buffer solutions

Buffer	pK$_a$
Sulfamic acid/sulfamate	1.0
Formic acid/formate	3.8
Acetic acid/acetate	4.7
Dihydrogenphosphate/hydrogenphosphate	7.1
Ammonium ammonia	9.3
Bicarbonate/carbonate	10.4
Fumaric acid/hydrogen fumarate	3.0
Benzoic acid/benzoate	4.2

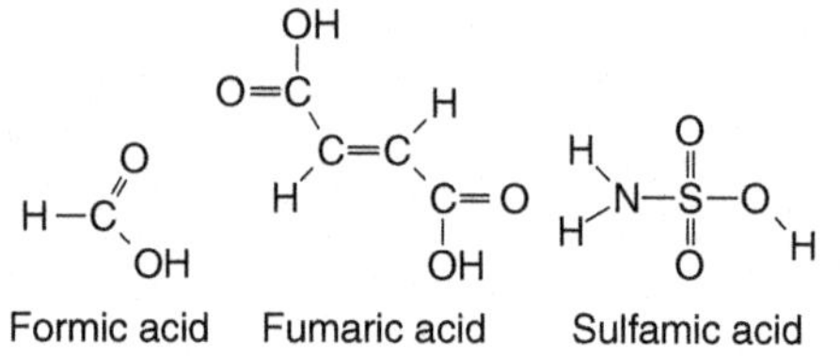

Formic acid   Fumaric acid   Sulfamic acid

Figure 33 Weak organic acids.

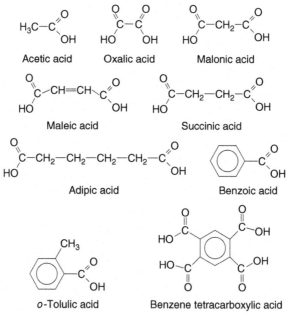

Acetic acid   Oxalic acid   Malonic acid

Maleic acid   Succinic acid

Adipic acid   Benzoic acid

o-Tolulic acid   Benzene tetracarboxylic acid

Figure 34 Carbonic and dicarbonic acids.

For example, an increased temperature stability of various gums can be achieved by adding sodium bicarbonate to the fracturing fluid and thus raising the pH of the fracturing fluid to 9.2-10.4.

3.5 Clay Stabilizers

Advances in clay-bearing formation treatment have led to the development of numerous clay stabilizing treatments and additives. Most additives used are high-molecular weight cationic organic polymers. However, it has been shown that these stabilizers are less effective in low-permeability formations [59].

3.5.1 Salts

The use of salts, such as potassium chloride and sodium chloride, as temporary clay stabilizers during oil well drilling, completion, and servicing, has been in practice for many years. Because of the bulk and potential environmental hazards associated with the salts, many operators have looked for alternatives to replace their use.

Recent research has developed a relationship between physical properties of various cations (e.g., K^+, Na^+) and their efficiency as temporary clay stabilizers. These properties were used to synthesize an organic cation (Table 42) with a higher efficiency as a clay stabilizer than the typical salts used in the oil industry to this point.

Table 42 Clay stabilizers

Compound	References
Ammonium chloride	
Potassium chloride[a]	[60]
Dimethyl diallyl ammonium salt[b]	[61]
N-Alkyl pyridinium halides	
N,N,N-Trialkylphenylammonium halides	
N,N,N-Trialkylbenzylammonium halides	
N,N-Dialkylmorpholinium halides[c]	[62, 63]
Reaction product of a homopolymer of MA and an alkyl diamine[d]	[64]
Tetramethylammonium chloride and methyl chloride	[65]
Quaternary salt of ethylene-ammonia condensation polymer[d]	
Quaternary ammonium compounds[e]	[66]

[a]Added to a gel concentrate with a diesel base.
[b]Minimum 0.05% to prevent swelling of clays.
[c]Alkyl equals methyl, ethyl, propyl, and butyl.
[d]Synergistically retards water absorption by the clay formation.
[e]Hydroxyl-substituted alkyl radials.

These additives provide additional benefits when used in conjunction with acidizing and fracturing treatments. A much lower salt concentration can be used to obtain the same clay stabilizing effectiveness [67, 68]. The liquid product has been proven to be much easier to handle and transport. It is environmentally compatible and biodegradable in its diluted form.

3.5.2 Amine Salts of Maleic Imide

Compositions containing amine salts of imides of MA polymers are useful for clay stabilization. These types of salts are formed, for example, by the

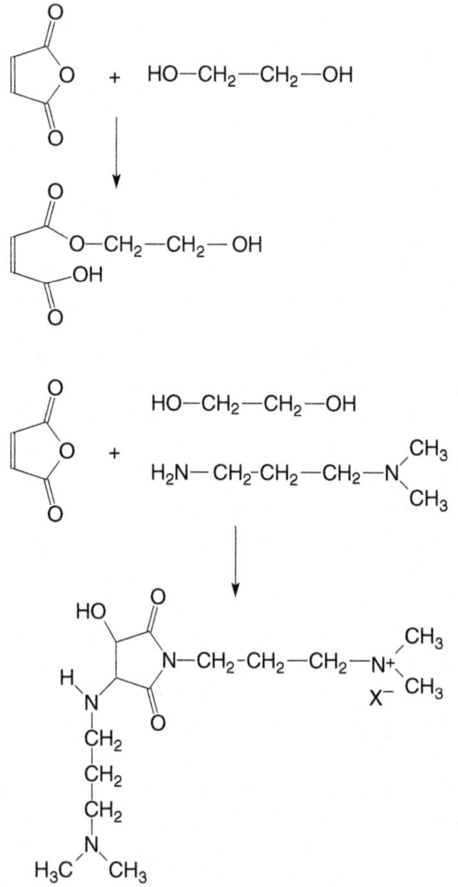

Figure 35 Start of condensation with ethylene glycol (top) and formation of amine salts of imides (bottom) [69].

reaction of MA with a diamine such as dimethyl aminopropylamine, in EG solution [69]. The primary nitrogen dimethyl aminopropylamine forms the imide bond.

In addition, it may add to the double bond of MA. Further, the EG may add to the double bond, but also may condense with the anhydride itself. On repetition of these reactions, oligomeric compounds may be formed. The elementary reactions are shown in Figure 35.

Finally, the product is neutralized with acetic acid or methane sulfonic acid to a pH of 4. The performance was tested in Bandera sandstone. The material neutralized with methane sulfonic acid is somewhat less than that neutralized with acetic acid. The compositions are particularly suitable for water-based hydraulic fracturing fluids.

3.6 Biocides

A hydraulic fracturing fluid containing guar gum or other natural polymers can be stabilized against bacterial attack by adding heterocyclic sulfur compounds. This method of stabilization prevents any undesired degradation of the fracturing fluid such as reduction of its rheologic properties (which are necessary for conducting the hydraulic fracturing operation) at high temperatures. Biocides suitable for fracturing fluids are shown in Table 43 and Figure 36.

Table 43 Biocides

Compound	References
Mercaptobenzimidazole[a]	[70]
1,3,4-Thiadiazole-2,5-dithiol[a,b]	[71–73]
2-Mercaptobenzothiazole	
2-Mercaptothiazoline	
2-Mercaptobenzoxazole	
2-Mercaptothiazoline	
2-Thioimidazolidone	
2-Thioimidazoline	
4-Ketothiazolidine-2-thiol	
N-Pyridineoxide-2-thiol	

[a]For guar gum.
[b]For xanthan gum.

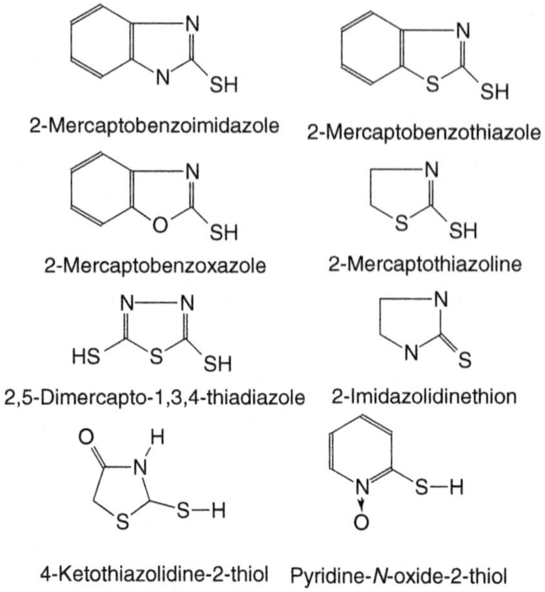

2-Mercaptobenzoimidazole   2-Mercaptobenzothiazole

2-Mercaptobenzoxazole   2-Mercaptothiazoline

2,5-Dimercapto-1,3,4-thiadiazole   2-Imidazolidinethion

4-Ketothiazolidine-2-thiol   Pyridine-*N*-oxide-2-thiol

Figure 36 Biocides for hydraulic fracturing fluids.

3.7 Surfactants

Surface active agents are included in most aqueous treating fluids to improve the compatibility of aqueous fluids with the hydrocarbon-containing reservoir. To achieve maximal conductivity of hydrocarbons from subterranean formations after fracture or other stimulation, it is the practice to cause the formation surfaces to be water-wet.

Alkylamino phosphonic acids and fluorinated alkylamino phosphonic acids adsorb onto solid surfaces, particularly onto surfaces of carbonate materials in subterranean hydrocarbon-containing formations, in a very thin layer. The layer is only one molecule thick and thus significantly thinner than a layer of water or a water-surfactant mixture on water-wetted surfaces [74–76].

These compounds so adsorbed resist or substantially reduce the wetting of the surfaces by water and hydrocarbons and provide high interfacial tensions between the surfaces and water and hydrocarbons. The hydrocarbons displace injected water, leaving a lower water saturation and an increased flow of hydrocarbons through capillaries and flow channels in the formation.

3.7.1 Viscoelastic Surfactants

Surfactants that are capable of developing viscoelasticity in aqueous solutions are of interest for a variety of wellbore fluids such as fracturing fluids [77, 78], selective fluids for water control [79, 80], drilling fluids, and scale dissolvers [81].

A wide range of VES solutions have been developed and formulated, with ionic headgroups including quaternary amines [82], amide and ester carboxylates [83], and amidesulfonates [84]. Also dimeric and oligomeric surfactants have been described [85]. The latter produce minimal emulsions when their aqueous solutions are mixed with hydrocarbons. In addition, VESs have been produced in a dry powder form as a convenient form of delivery [86].

The reaction between ethanol amines and fatty acids was first reported in 1923 [87]. Ethanol amine stearates have been used in a variety of industrial and medical applications [88–93].

A VES solution based on non-ionic amidoamine oxides for use in wellbore service fluids has been described [94, 95].

On the other hand, relatively insoluble carboxylates, for example, unbranched, long chain fatty acids with iodine values of less than 40, such as stearic acid and partially hydrogenated tallow acids, can be sufficiently solubilized at ambient temperatures by ethanol amines, eventually to generate viscoelastic solutions [96]. Typically, these surfactant solutions produced are cloudy and viscous at ambient temperatures but clarify and further viscosify on heating. They can reach a maximum viscosity above ambient temperatures. Suitable combinations of acids and amines are shown in Table 44.

Alkylol amines are shown in Figure 37.

Table 44 Compounds for alkylol amine salts of fatty acids [96]

Acid compound	Amine compound
Behenic acid	Ethanol amine
Behenic acid	Triethanol amine
Oleic acid	Ethanol amine
Stearic acid	Diethanol amine
Stearic acid	Ethanol amine
Tallow acid	Ethanol amine

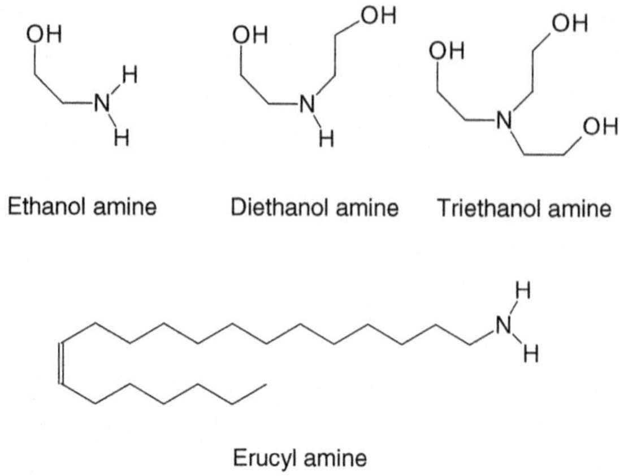

Ethanol amine   Diethanol amine   Triethanol amine

Erucyl amine

Figure 37 Alkylol amines.

A methyl quaternized erucyl amine is useful for aqueous viscoelastic Surfactant-based fracturing fluids in high-temperature and high-permeability formations [97].

3.8 Crosslinking Agents

3.8.1 Kinetics of Crosslinking

The rheology of hydroxypropyl guar is greatly complicated by the crosslinking reactions with titanium ions. A study to better understand the rheology of the reaction of hydroxypropyl guar with titanium chelates and how the rheology depends on the residence time, shear history, and chemical composition has been performed [98].

Rheologic experiments were performed to obtain information about the kinetics of crosslinking in hydroxypropyl guar. Continuous flow and dynamic data suggest a crosslinking reaction order of approximately 4/3 and 2/3, respectively, with respect to the crosslinking agent and hydroxypropyl guar concentration. Dynamic tests have shown that the shearing time is important in determining the final gel properties.

Continued steady shear and dynamic tests show that high shear irreversibly destroys the gel structure, and the extent of the crosslinking reaction decreases with increasing shear. Studies at shear rates below 100 s^{-1} suggest a shear-induced structural change in the polymer that affects the chemistry of the reaction and the nature of the product molecule.

3.8.2 Stability of Crosslinking Agents

A method for increasing the stability of water-based fracturing fluids is shown in which a base fracturing fluid is formulated by mixing together a hard mix water containing multivalent cations, a water-soluble polymer and a crosslinking agent for the water-soluble polymer. A water softener is added to the base fracturing fluid which hinders the ability of the multivalent cations present in the hard mix water to compete with the water-soluble polymer for borate-based crosslinking agent, thereby stabilizing the resulting fracturing fluid.

Traditional complexing agents used in the industry are borates and the transition metal-oxy complexes such as zirconates and the titanates. While such crosslinking agents or complexing agents are relatively non-problematic in freshwater, stability problems arise when seawater, brine, or hard water is used as the mix water.

One early attempt to overcome the above problem involved the precipitation of the multivalent cations in the mix water as metal carbonates. This procedure, while increasing the stability of the fluid, was less than desirable because the precipitated carbonates tended to cause formation damage and, at the very least, required an acid treatment to re-stimulate production.

A need exists for a method for stimulating a subterranean well using a water-based fracturing fluid which is complexed or crosslinked with commonly available agents and which is not affected by the use of hard mix water such as seawater.

A need also exists for such a method which provides a stable fracturing fluid having adequate proppant transport capabilities and viscosity to provide adequate fractured geometries over a broad temperature range [99].

3.8.3 Delayed Crosslinking

Delayed crosslinking is desirable because the fluid can be pumped down more easily. A delay is a retarded reaction rate of crosslinking. This can be achieved with the methods explained in the following section.

3.8.4 Borate Systems

Boric acid can form complexes with hydroxyl compounds. The mechanism of formation complexes of boric acid with glycerol is shown in Figure 38. Three hydroxyl units form an ester and one unit forms a complex bond. Here, a proton will be released that lowers the pH. The scheme is valid also for polyhydroxy compounds. In this case, two polymer chains are connected via such a link.

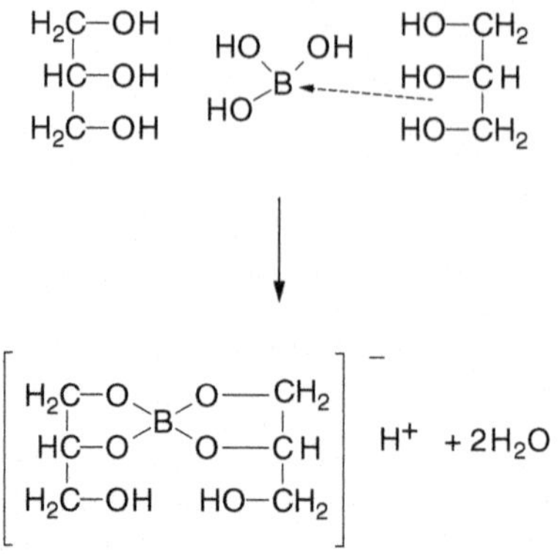

Figure 38 Complexes of boric acid with glycerol.

The control of the delay time requires the control of the pH, the availability of borate ions, or both. Control of pH can be effective in freshwater systems [100]. However, the control of borate is effective in both fresh water and sea water. This may be accomplished by using sparingly soluble borate species or by complexing the borate with a variety of organic species.

Borate-crosslinked fracturing fluids have been successfully used in fracturing operations. These fluids provide excellent rheologic, fluid loss, and fracture conductivity properties over fluid temperatures up to 105 °C. The mechanism of borate crosslinking is an equilibrium process that can produce very high fluid viscosities under conditions of low shear [101]. A fracturing fluid containing borate is prepared in the following way [102]:

1. introducing a poly(saccharide) polymer into (sea) water to produce a gel;
2. adding an alkaline agent to the gel to obtain a pH of at least 9.5; and
3. adding a borate crosslinking agent to the gel to crosslink the polymer.

A dry granular composition can be prepared in the following way [103]:

1. dissolving from 0.2% to 1.0% of a water-soluble poly(saccharide) in an aqueous solution;
2. admixing a borate source with the aqueous gel formed in Step 1;
3. drying the borate-crosslinked poly(saccharide) formed in Step 2; and
4. granulating the product of Step 3.

A borate crosslinking agent can be boric acid, borax, an alkaline earth metal borate, or an alkali metal alkaline earth metal borate. The borate source, calculated as boric oxide, must be present in an amount of 5-30%.

Borated starch compositions are useful for controlling the rate of crosslinking of hydratable polymers in aqueous media for use in fracturing fluids. The borated starch compositions are prepared by reacting, in an aqueous medium, starch and a borate source to form a borated starch complex. This complex provides a source of borate ions, which cause crosslinking of hydratable polymers in aqueous media [104]. Delayed crosslinking takes place at low temperatures.

Efficient crosslinkers for fluids with reduced polymer loading are of great interest, to reduce both the formation and the proppant-pack damage from polymer residues. The synthesis of poly(aminoboronate)s has been detailed [105]. This method was improved by using a readily available polyamine as the base scaffold, and boron was incorporated during the condensation reaction between boric acid and EG. In this way, the volatile and highly flammable trimethylborate was eliminated [106]. The synthesis is shown in Figure 39.

Glyoxal [107, 108] is effective as a delay additive within a certain pH range. Glyoxal is shown in Figure 40. It bonds chemically with both boric acid and the borate ions to limit the number of borate ions initially available in solution for subsequent crosslinking of a hydratable poly(saccharide) (e.g., galactomannan).

The subsequent rate of crosslinking of the poly(saccharide) can be controlled by adjusting the pH of the solution. The mechanism of delayed crosslinking is shown in Figure 41. If two hydroxyl compounds with low molecular weight are exchanged with high-molecular weight compounds, the hydroxyl units belonging to different molecules, then a crosslink is formed.

Other dialdehydes, keto aldehydes, hydroxyl aldehydes, ortho-substituted aromatic dialdehydes, and ortho-substituted aromatic hydroxyl aldehydes have been claimed to be active in a similar way [109]. Borate-crosslinked guar fracturing fluids have been reformulated to allow use at high temperatures in both fresh water and sea water.

The temporary temperature range is extended for the use of magnesium oxide-delayed borate crosslinking of a galactomannan gum fracturing fluid by adding fluoride ions that precipitate insoluble magnesium fluoride [110].

Alternatively, a chelating agent for the magnesium ion may be added. With the precipitation of magnesium fluoride or the chelation of the magnesium ion, insoluble magnesium hydroxide cannot form at elevated temperatures, which would otherwise lower the pH and reverse the borate crosslinking reaction. The addition effectively extends the use of such fracturing fluids to temperatures of 135-150 °C.

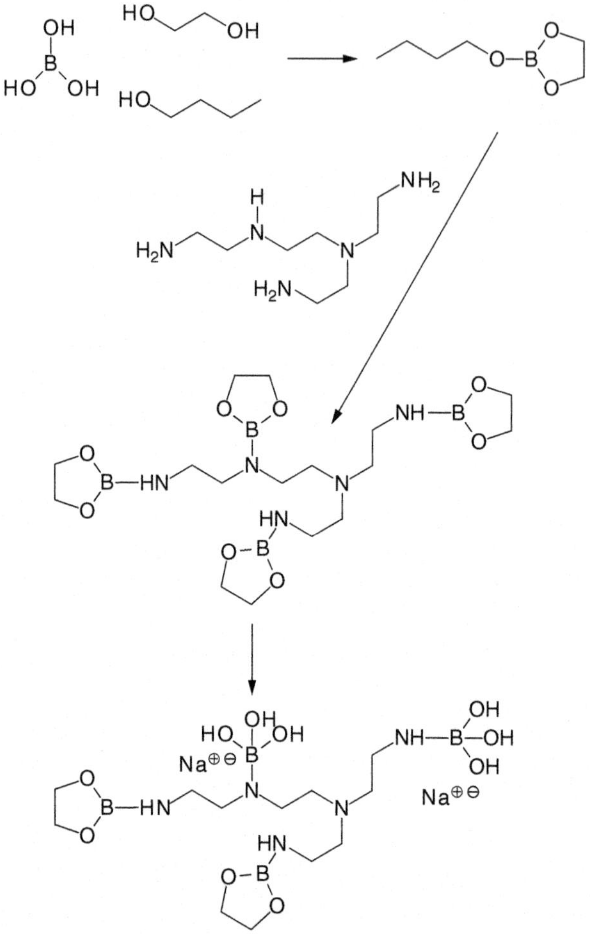

Figure 39 Synthesis of a borate crosslinking agent [106].

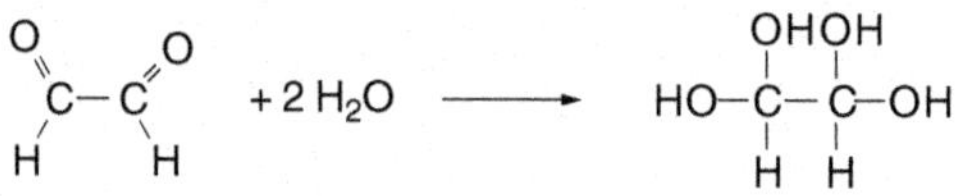

Figure 40 Glyoxal and hydrate formation.

Polyols, such as glycols or glycerol, can delay the crosslinking of borate in hydraulic fracturing fluids based on galactomannan gum [111]. This is suitable for high-temperature applications up to 150 °C. In this case, low-molecular weight borate complexes initially are formed but exchange slowly with the hydroxyl groups of the gum.

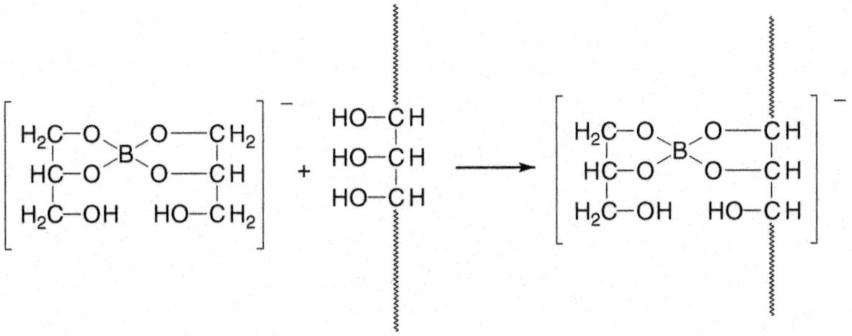

Figure 41 Delayed crosslinking.

3.8.5 Titanium Compounds

Organic titanium compounds are useful as crosslinking agents [112]. Aqueous titanium compositions often consist of mixtures of titanium compounds.

3.8.6 Zirconium Compounds

Various zirconium compounds are used as delayed crosslinking agents, c.f., Table 45. The initially formed complexes with low–molecular weight compounds are exchanged with intermolecular poly(saccharide) complexes, which cause delayed crosslinking.

A diamine-based compound for complex forming is shown in Figure 42.

Hydroxy acids are shown in Figure 43. Polyhydroxy compounds suitable for complex formation with zirconium compounds are shown in Figure 44.

Borozirconate complexes can be prepared by the reaction of tetra-*n*-propyl zirconate with triethanol amine and boric acid [119]. The borozirconate complex can be used at a pH of 8-11.

Table 45 Zirconium compounds suitable as delayed crosslinking agents

Zirconium crosslinking agents/chelate	References
Hydroxyethyl-tris-(hydroxypropyl) ethylene diamine[a]	[113]
Zirconium halide chelates	[114]
Boron zirconium chelates[b]	[115–118]

[a] Good high-temperature stability.
[b] High-temperature application, enhanced stability.

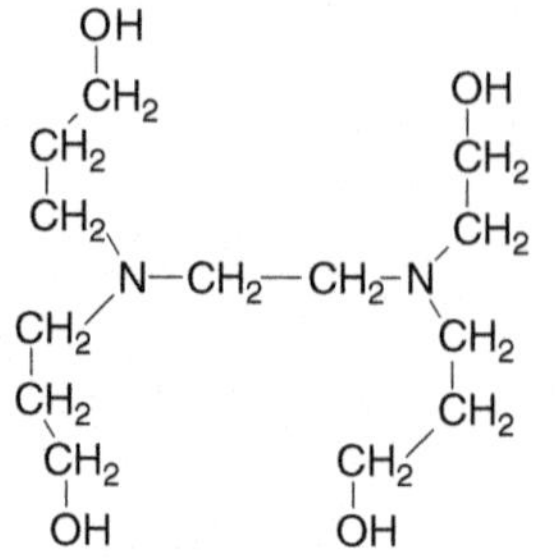

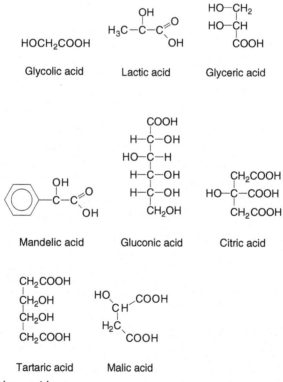

Figure 42 Hydroxyethyl-tris-(hydroxypropyl) ethylene diamine.

HOCH₂COOH

Glycolic acid

Lactic acid

Glyceric acid

Mandelic acid

Gluconic acid

Citric acid

Tartaric acid

Malic acid

Figure 43 Hydroxy acids.

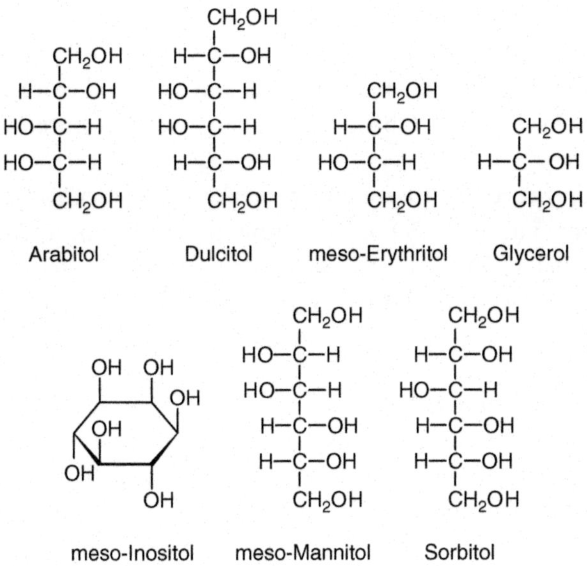

Figure 44 Polyalcohols for complex formation.

3.9 Gel Breaking in Water-Based Systems

In general, there are two methods for combining the fracturing fluid and the breaker [120]:

1. mixing the breaker with the fracturing fluid prior to sending the fracturing fluid downhole, or
2. sending the fracturing downhole, and afterwards the breaker.

The first method is favored at least because of convenience. It is easier to mix the fluids at the surface and send the mixture downhole. A disadvantage of this blending method is that the breaker can decrease the viscosity of the fracturing fluid before the desired time.

In the second method, the fracturing fluid is sent downhole, and the breaker is sent downhole later. While sending the breaker downhole later is inconvenient, in this method, the breaker does not decrease the viscosity of the fracturing fluid prematurely [120].

After the fracturing job, the properties of the formation should be restored. Maximal well production can be achieved only when the solution viscosity and the molecular weight of the gelling agent are significantly reduced after the treatment, that is, the fluid is degraded.

3.9.1 Basic Studies

Comprehensive research on the degradation kinetics of a hydroxypropyl guar fracturing fluid by enzyme, oxidative, and catalyzed oxidative breakers was performed [121–123]. Changes in viscosity were measured as a function of time.

The studies revealed that enzyme breakers are effective only in acid media at temperatures of 60 °C or below. In an alkaline medium and at temperatures below 50 °C, a catalyzed oxidative breaker system was the most effective breaker. At temperatures of 50 °C or higher, hydroxypropyl guar fracturing fluids can be degraded by an oxidative breaker without a catalyst.

3.9.2 Oxidative Breakers

Among the oxidative breakers, alkali, metal hypochlorites, and inorganic and organic peroxides have been described in literature. These materials degrade the polymer chains by oxidative mechanisms. CMC, guar gum, or partially hydrolyzed poly(acrylamide) (PHPAs) were used for testing a series of oxidative gel breakers in a laboratory study [124]. Water–soluble polymer breakers are summarized in Table 46.

Hypochlorite Salts

Hypochlorites are powerful oxidants and therefore may degrade polymeric chains. They are often used in combination with tertiary amines [126]. The combination of the salt and the tertiary amine increases the reaction

Table 46 Water-soluble polymer breakers [125]

Compound
Ammonium persulfate
Sodium persulfate
Potassium persulfate
Sodium peroxide
Barium peroxide
Hydrogen peroxide
Magnesium peroxide
Potassium peroxide
Sodium perborate
Potassium perborate
Potassium permanganate
Sodium permanganate

rate more than the application of a hypochlorite alone. A tertiary amino galactomannan may serve as an amine source [127].

This also serves as a thickener before breaking. Hypochlorites are also effective for breaking stabilized fluids [128]. Sodium thiosulfate has been proposed as a stabilizer for high-temperature applications.

Peroxide Breakers

Alkaline earth metal peroxides have been described as delayed gel breakers in alkaline aqueous fluids containing hydroxypropyl guar [129]. The peroxides are activated by increasing the temperature of the fluid.

Perphosphate esters or amides can be used for oxidative gel breaking [130]. Whereas the salts of the perphosphate ion interfere with the action of the crosslinking agents, the esters and amides of perphosphate do not.

Fracturing fluids that contain these breakers are useful for fracturing deeper wells operating at temperatures of 90-120 °C and using metal ion crosslinking agents such as titanium and zirconium. Breaker systems based on persulfates have also been described [131].

In addition, organic peroxides are suitable for gel breaking [132]. The peroxides need not be completely soluble in water. The time needed to break is controlled in the range of 4-24 h by adjusting the amount of breaker added to the fluid.

3.9.3 Redox Gel Breakers

Basically, gel breakers act according to a redox reaction. Copper(II) ions and amines can degrade various poly(saccharide)s [133].

3.9.4 Delayed Release of Acid

Regained permeability studies with HEC polymer in high-permeability cores revealed that persulfate-type oxidizing breakers and enzyme breakers do not adequately degrade the polymer. Sodium persulfate breakers were found to be thermally decomposed, and the decomposition was accelerated by minerals present in the formation.

The enzyme breaker adsorbed onto the formation but still partly functioned as a breaker. Dynamic fluid loss tests at reduced pH with borate-crosslinked gels suggest that accelerated leak-off away from the wellbore could be obtained through the use of a delayed release acid. Rheologic measurements confirmed that a soluble delayed release acid could be used to convert a borate-crosslinked fluid into a linear gel [134].

Figure 45 Hydrolysis of poly(glycolic acid).

A condensation product of hydroxyacetic acid can be used as a fluid loss material in a fracturing fluid in which another hydrolyzable aqueous gel is used [135–138]. The hydroxyacetic acid condensation product degrades at formation conditions to set free hydroxyacetic acid, which breaks the aqueous gel. This mechanism may be used for delayed gel breaking, as shown in Figure 45. Here, the permeability is restored without the need for separate addition of a gel breaker, and the condensation product acts a fluid loss additive.

3.9.5 Enzyme Gel Breakers

Enzymes specifically cleave the backbone structure of the thickeners and eventually of the fluid loss additive. They offer several advantages to other breaker systems because of their inherent specificity and the infinite polymer-degrading activity. Initially the application of enzymes has been limited to low-temperature fracturing treatments because of pH and temperature constraints. Only recently, extreme temperature-stable and polymer-specific enzymes have been developed [139].

Basic Studies

Basic studies have been performed to investigate the performance of enzymes. The products of degradation, the kinetics of degradation, and limits of application, such as temperature and pH, have been analyzed [140, 141]. Because enzymes degrade chemical linkages highly selectively, no general purpose enzyme exists, but for each thickener, a selected enzyme must be applied to guarantee success. Enzymes suitable for particulate systems are shown in Table 47.

Enzymes are suitable to break the chains of the thickener directly. Other systems also have been described that enzymatically degrade polymers,

Table 47 Polymer enzyme systems

Polymer	References
Xanthan[a]	[142]
Mannan-containing hemicellulose[b]	[143]

[a]Elevated temperatures and salt concentrations.
[b]High alkalinity and elevated temperature.

which degrade into organic acid molecules. These molecules are actually active in the degradation of the thickener [144].

Interactions

Despite their advantages over conventional oxidative breakers, enzyme breakers have limitations because of interferences and incompatibilities with other additives. Interactions between enzyme breakers and fracturing fluid additives including biocides, clay stabilizers, and certain types of resin-coated proppants have been reported [145].

3.9.6 Encapsulated Gel Breakers

The breaker chemical in encapsulated gel breakers is encapsulated in a membrane that is not permeable or is only slightly permeable to the breaker. Therefore, the breaker may not come in contact initially with the polymer to be degraded. Only with time the breaker diffuses out from the capsulation, or the capsulation is destroyed so that the breaker can act successfully.

Encapsulated gel breakers find a wide field of application for delayed gel breaking. The breaker is prepared by encapsulating it with a water-resistant coating. The coating shields the fluid from the breaker so that a high concentration of breaker can be added to the fluid without causing premature loss of fluid properties such as viscosity or fluid loss control.

Critical factors in the design of encapsulated breakers are the barrier properties of the coating, release mechanisms, and the properties of the reactive chemicals. For example, a hydrolytically degradable polymer can be used as the membrane [146].

This method of delayed gel breaking has been reported both for oxidative breaking and for enzyme gel breaking. Formulations of encapsulated gel breakers are shown in Table 48. Membranes for encapsulators are shown in Table 49.

Table 48 Use of encapsulation in delayed gel breaking

Breaker system	References
Ammonium persulfate[a]	[147–150]
Enzyme breaker[b]	[151]
Complexing agents[c]	[152]

[a]Guar or cellulose derivatives.
[b]Open cellular coating.
[c]For titanium and zirconium; wood resin encapsulated.

Table 49 Membranes for encapsulated breakers

Membrane material	References
PA[a]	[153, 153]
Crosslinked elastomer	[154]
Partially hydrolyzed acrylics crosslinked with aziridine prepolymer or carbodiimide[b]	[155–157]
7% Asphalt and 93% neutralized sulfonated ionomer	[158]

[a]For peroxide particle sizes 50-420 μm.
[b]Enzyme coated on cellulose derivative.

3.9.7 Gel Breaking of Guar

Maximal well production can be achieved only when the solution viscosity and the molecular weight of the gelling agent are significantly reduced after the treatment. However, the reduction of the fracturing fluid viscosity, the traditional method of evaluating these materials, does not necessarily indicate that the gelling agent has been thoroughly degraded also.

The reaction between hydroxypropyl guar and the oxidizing agent (ammonium peroxydisulfate) in an aqueous potassium chloride solution was studied [159] under controlled conditions to determine changes in solution viscosity and the weight average of the molecular mass of hydroxypropyl guar.

Bromine compositions used for gel breaking can be stabilized with sodium sulfamate [120]. The sulfamate used in the production of such breakers is effective in stabilizing the active bromine species over long periods of time, especially at a pH of 13. For example, a WELLGUARD™ 7137 gel breaker is stable for greater than 1 year if protected from sunlight. The halogen source of the breaker are interhalogen compounds, bromine chloride, or mixtures of bromine and chlorine.

Unlike hypobromites ($^-$OBr), these type breakers do not oxidize or otherwise destroy organic phosphonates that are typically used as corrosion and scale inhibitors. Further, the breakers exhibit a low corrosivity against metals, especially against ferrous alloys. This is the result of the low oxidation-reduction potential of these breakers [120, 160]. The effect of the breakers on guar is shown in Figure 46. The composition was prepared and studied at 50 °C (120 °F).

Borate-crosslinked guar polymer gels can be broken with EDTA compounds [161]. EDTA and other aminocarboxylic acid compounds can break the gelled fracturing fluid. Examples are shown in Table 50 and Figure 47.

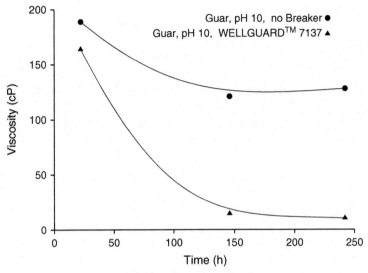

Figure 46 Effect of halogen-based breakers on guar [120].

Table 50 EDTA-related gel breakers [161]
Complex compound

Tetrasodium propylene diamine tetraacetic acid
Trisodium hydroxyethylene diamine tetraacetic acid
Trisodium nitrilo triacetic acid
Trisodium ethylene diamine triacetic acid
Disodium ethylene diamine diacetic acid
Disodium calcium dihydrate ethylene diamine diacetic acid
Tetraammonium ethylene diamine tetraacetic acid

It is believed that these breakers act directly on the polymer itself and not on any crosslinking agent that may be present. Also polyhydroxy compounds can break guar gels, and moreover gels formed by poly(saccharide)s. These polyhydroxy compounds include mannitol and sorbitol. The polyols can be used in combination of enzyme breakers [162].

3.9.8 Gel Breaking of Viscoelastic Surfactant-Gelled Fluids

The viscosity of fluids that are viscosified with VESs can be controlled by fatty acid salts. For example, a brine fluid gelled with an amine oxide surfactant may have its viscosity broken with a composition containing naturally occurring fatty acid salts from canola oil or corn oil [163].

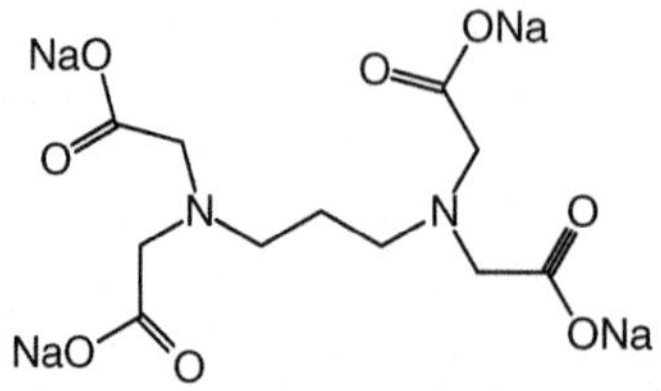

Tetrasodium propylene diamine tetraacetic acid

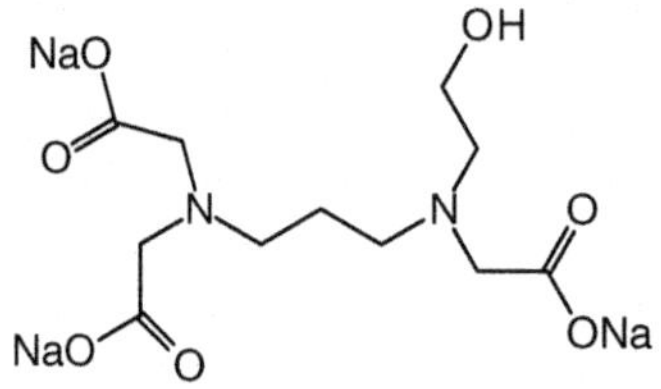

Trisodium hydroxyethylene diamine tetraacetic acid

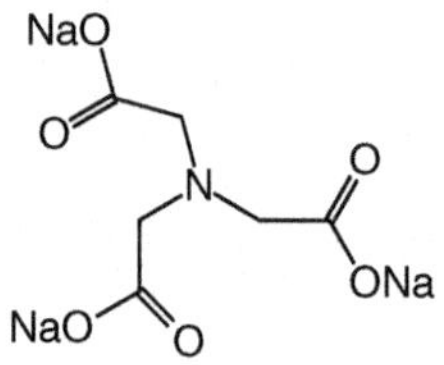

Trisodium nitrilo triacetic acid

Figure 47 EDTA-related gel breakers.

The alteration of the fatty acid or the saponification reaction may occur during mixing and pumping of the fluid downhole. The method may also be used where most of the saponification occurs within the reservoir shortly after the treatment is over. Alternatively, the components may be preformed and added later as an external breaker solution to remove the VES–gelled fluids that have been already placed downhole.

It may be possible that first an increase in viscosity and afterwards a decrease in viscosity may occur. When canola oil is saponified with CaOH, initially a slight increase of the VES fluid is observed, followed by a breaking reaction [163]. The increase in viscosity is explained, as the initially particular saponified fatty acids may act as viscosity enhancing cosurfactants for the fluid containing VESs.

3.9.9 Granules

Granules may also be helpful in delayed breaking. Granules with 40-90% of sodium or ammonium persulfate breaker and 10-60% of an inorganic powdered binder, such as clay, have been described [164]. The granules exhibit a delayed release of the breaker.

Other chemicals acting as delayed gel breakers are also addressed as controlled solubility compounds or cleanup additives that slowly release certain salts. Poly(phosphate)s have been described as such [165].

Granules composed of a particulate breaker dispersed in a wax matrix are used in fracturing operations to break hydrocarbon liquids gelled with salts of alkyl phosphate esters. The wax granules are solid at surface temperature and melt or disperse in the hydrocarbon liquid at formation temperature, releasing the breaker to react with the gelling agent [166].

3.10 Scale Inhibitors

The formation of calcium carbonate ($CaCO_3$), calcium sulfate ($CaSO_4$), and barium sulfate ($BaSO_4$) scales in brine may create problems with permeability. Therefore, it is advantageous that newly made fractures have a scale inhibitor in place in the fracture to help prevent the formation of scale. Formulations of hydraulic fracturing fluids containing a scale inhibitor have been described in the literature [167].

3.10.1 Interference of Chelate Formers

Trace amounts of metal chelate-forming additives, which are used in fracture fluids, have been shown to have a debilitating effect on the performance of widely used barium sulfate scale inhibitors. Ethylene diamine tetraacetic acid, citric acid, and gluconic acid render some scale inhibitors, such as phosphonates, polycarboxylates, and phosphate esters, completely ineffective at concentrations as low as $0.1 \, \mathrm{mg\,l^{-1}}$. Such low concentrations may be expected to return from formation stimulation treatments for many months and would appear to jeopardize any scale inhibitor program in place.

This conclusion follows from experiments with a simulated North Sea scaling system at pH 4 and 6. The scale inhibitor concentrations studied were 50 and 100 $\mathrm{mg\,l^{-1}}$. The large negative effect of the organic chelating agents was observed at pH 4 and 6. The only scale inhibitors studied that remained unaffected by these interferences were poly(vinyl sulfonate)s [168].

3.10.2 Encapsulated Scale Inhibitors

A solid, encapsulated scale inhibitor (calcium-magnesium poly(phosphate)) has been developed and extensively tested for use in fracturing treatments [169–171]. The inhibitor is compatible with borate-crosslinked and zirconium-crosslinked fracturing fluids and foamed fluids because of coating.

The coating exhibits a short-term effect on the release rate profile. The composition of the solid derivative has the greatest effect on its long-term release rate profile.

3.10.3 Thermal Degradation

The thermodynamics and kinetics of the thermal degradation of scale inhibitors has been elucidated [172].

An efficient testing approach based on inhibition kinetics has been developed and successfully applied to determine the fraction of the active inhibitor molecules in preheated samples of scale inhibitors with various generic chemistries. The kinetics of the thermal degradation has been modeled based on the integrated first-order rate equation and Arrhenius equation. The thermal degradation of the inhibitors was analyzed by nuclear-magnetic-resonance (NMR) spectroscopy.

4. SPECIAL ISSUES OF WATER-BASED FRACTURING FLUIDS

4.1 Water Mixing During Waterflood Oil Recovery

The mixing of injected water and in situ water during waterfloods has been investigated [173]. It turned out that the mixing process is sensitive to the initial water saturation. Differences between a waterflooded zone and a preflooded zone have been elucidated during a water-based EOR displacement process.

The mixing of in situ, or connate, water and injected water during laboratory waterfloods in a strongly water-wet chalk core sample was assessed at different initial water saturations.

Nuclear-tracer imaging was used during the waterfloods, in order to distinguish the water between the oil phase, connate water, and injected water.

The mixing of connate and injected water during a waterflood experiment, in the presence of an oil phase, resulted in a displacement of all connate water from the core plug.

During the displacement, connate water banked in front of the injecting water is separating the injected water from the mobile oil phase.

The effect of the connate water bank separation is sensitive to the initial water saturation. The time difference between the breakthrough of the connate water and the breakthrough of the injected water at the outlet shows a linear correlation to the initial water saturation [173].

4.2 Displacement of Connate Water

The movement of connate water was studied in laboratory waterflooding experiments of oil-saturated chalk at connate water saturation from a North Sea reservoir [174]. The water was labeled with spiked with ^{22}Na, a radioactive sodium isotope.

Using a γ-monitoring technique, it was found that the connate water is piled up at the front of the injected water and a mixed water bank with almost 100% connate water in the front is formed. Behind this region, there is a gradual transition to pure injection water.

4.3 Effect of Viscous Forces and Initial Water Saturation

The effect of wettability and initial water saturation on water injection and imbibition was made in Kansas outcrop chalk samples [175]. The initial water saturation has a very pronounced effect on water injection in an intermediate-wet chalk. In contrast, this effect is much less in a strongly water-wet chalk. In addition, the viscous forces have a strong effect on the performance of a water injection operation in intermediate-wet chalks. The results from field experiments and laboratory measurements indicate no correlation between spontaneous imbibition and field performance.

4.4 Waterflooding Performance in Horizontal Wells

The factors that are determining the performance of waterflooding in horizontal wells are heterogeneity, horizontal well length, location, and trajectory [176].

From the production data of horizontal wells with bottom water using a numerical simulation for production history matching the waterflooded patterns of horizontal wells with bottom water reservoir were classified into three categories [176]:

1. breakthrough waterflooded the whole horizontal section;
2. punctiform breakthrough waterflooded the whole horizontal section; and

3. punctiform breakthrough waterflooded the local horizontal section.

From these results a qualitative design of a horizontal well, for the completion plan was established, including water shutoff, hydraulic fracturing, or acidizing [176].

4.5 Recycling Water: Case Studies in Designing Fracturing Fluids

In general, the flowback water in well stimulation contains the chemicals and/or byproducts of the hydraulic fracturing process [177].

The produced water is a different from natural formation brine that is produced along with the hydrocarbons from the well, and may contain large quantities of dissolved salts, dispersed hydrocarbons, etc. These waters are considered waste byproducts of oil and gas production can make problems with regard to disposal. If the flowback and produced water can be recycled, the total amount of fresh water that must be used in the process can be reduced.

Specific and customized fluid compositions have been developed that can be used effectively with recycled waters. Also successful case studies have been presented [177].

4.6 Influence of Water Injection and Natural Water

It has been shown that the influence of water injection and natural water can be discriminated effectively in an Jurassic sandstone reservoir [178]. A new method based on the principle of material balance has been presented to assess the discrimination.

The injected water and the influx water are subdivided into two parts according to their function in the reservoir: One part that is maintaining the reservoir energy, and the other is effective as driving fluid. On the other hand, the production fluids are formed due to two factors: One is the driving of the injection water and the influx water, and the other is due to the depletion of the reservoir energy.

The contributions of the water injection and the natural water influx to the reservoir pressure system have been calculated using the actual production history. These results can be used for the adjustment and optimization of the future production [178].

4.7 Water Coning

Water coning may become a serious problem in mature reservoirs where the bottom water drive overlays depleted oil strata [179]. High values of water cut and minimal oil rate may result in an early shut down of the wells without a sufficient recovery of the hydrocarbons in place.

The downhole water sink technology is a comparatively new and effective method to control the water coning [179]. Theoretical studies and field practice have shown that the downhole water sink technology can improve oil production rate, reduce water cut, and increase oil productivity index of a well. The cost of production increases as this technique requires draining and lifting of large amounts of water from the aquifer.

So, another water coning control method has been developed, the downhole water loop installation. This method has the same advantages as the downhole water sink technology technique does not need to lift the water to the surface. The feasibility of the downhole water loop installation technique has been studied with a simple analytical model [179]. The model matches well to real production data.

4.8 Fracturing of Carbonate Formations

For carbonate formations, a reactive fluid, such as acid is pumped through the tubing side and at the same time a nonreactive fluid is pumped through the annular space. In the case of a carbonate formation, a water-based fluid may be the nonreactive fluid. Both fluids are mixed downhole and are responsible for the fracture creation or enhancement [1].

In some circumstances, the fracturing process may terminate prematurely. So, the nonreactive portion of the stimulation fluid, which usually is intended to advance as the fracture progresses, may undesirably completely leak-off into the formation and result in an inefficient fracture stimulation of the well.

In fracturing treatments, a fluid loss into the formation results in a reduction in efficiency as the fracturing fluid cannot propagate the fractures as desired. This drawback can be overcome by the use of viscosified or crosslinked fluids.

Also, a relative permeability modifier can be added to reduce the fluid leak-off. A relative permeability modifier refers to a compound that is capable of reducing the permeability of a subterranean formation to aqueous-based fluids without substantially changing its permeability to hydrocarbons [180].

This is a water-soluble polymer with a hydrophobic modification. Importantly, hydrophobically modified polymers that have hydrophobic groups incorporated into the hydrophilic polymer structure should remain water soluble [1].

Examples of hydrophilic base polymers are poly(acrylamides), poly(vinylamine)s, and copolymers of vinylamines and vinyl alcohols. These hydrophilic base polymers can be reacted with hydrophobic compounds such as alkyl halides, sulfonates, sulfates, organic acids, and organic acid derivatives. Specific examples are given in Table 51 and Figure 48.

Table 51 Hydrophobic compounds [1]

Compounds for reaction with hydrophilic polymers

Octenyl succinic acid
Dodecenyl succinic acid
Octenyl succinic acid ester
Dodecenyl succinic acid ester
Octenyl succinic acid amide
Dodecenyl succinic acid amide

Monomers for copolymerization with hydrophilic monomers

Octadecyldimethylammoniumethyl methacrylate bromide
Hexadecyldimethylammoniumethyl methacrylate bromide
Hexadecyldimethylammoniumpropyl methacrylamide bromide
2-Ethylhexyl methacrylate
Hexadecyl methacrylamide

Octenyl succinic acid   Hexadecyl methacrylamide

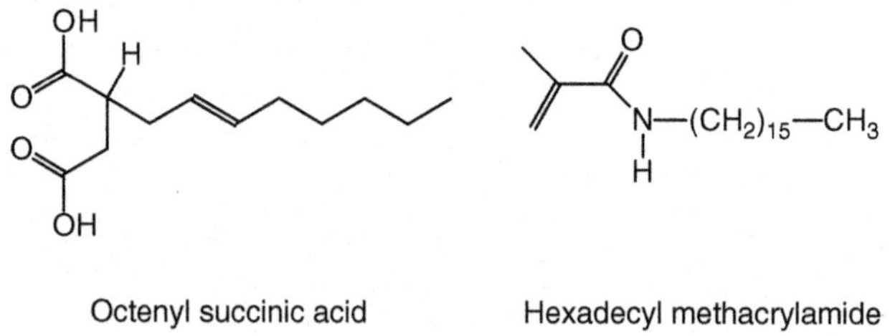

Figure 48 Hydrophobic monomers.

Examples of methods of the preparation of such modified polymer have been described [181]. The polymerization reaction can be performed in various ways. An example of a particularly suitable procedure for polymerizing water-soluble monomers is as follows:

Preparation III–1: Into a 250-ml three-neck round bottom flask, the following is charged: 47.7 g distilled water, 1.1 g acrylamide, and 0.38 g alkyl dimethylammoniumethyl methacrylate bromide. The solution formed is sparged with nitrogen for approximately 30 min followed by the addition of 0.0127 g of 2,2′-azobis(2-amidinopropane) dihydrochloride. The resulting solution is then heated, with stirring, to 45 °C and held for 18 h to produce a highly viscous polymer solution. ■

When the hydrophobically modified hydrophilic monomer is not water soluble, for example, octadecyl methacrylate, the following procedure can be used:

Preparation III–2: Into a 250-ml three-neck round bottom flask, the following is charged: 41.2 g distilled water and 1.26 g acrylamide. The solution formed is sparged with nitrogen for approximately 30 min, followed by the addition of 0.06 g of octadecyl methacrylate and 0.45 g of a cocamidopropyl betaine surfactant. The mixture is stirred until a homogeneous, clear solution is obtained followed by the addition of 0.0055 g of 2,2′-azobis(2-amidinopropane) dihydrochloride. The resulting solution is then heated, with stirring, to 45 °C and held for 18 h to produce a highly viscous polymer solution. ■

4.8.1 Lactic Acid

Lactic acid has been used in the oil field for acid fracturing and filter cake removal, further, as an iron-control agent during acid treatments. The determination of the reaction rate and the acid diffusion properties is a critical step for successful treatments in matrix acidizing and acid fracturing [182]. Mass transfer and reaction kinetics have been reported for the lactic acid/calcite system using a rotating-disk apparatus.

At 27 °C, the reaction of lactic acid with calcite is controlled by mass transfer at low disk rotational speeds and is surface reaction limited at higher speeds. At high temperatures of 55-120 °C, both mass transfer and surface reaction influence the overall calcite dissolution [182].

Tradenames appearing in the references are shown in Table 52.

Table 52 Tradenames in references
Tradename

Description	Supplier
Alkaquat™ DMB-451 Dimethyl benzyl alkyl ammonium chloride [37]	Rhodia Canada, Inc.
Ammonyx® GA-90 Dipalmitoylethyl hydroxyethylmonium methosulfate [83]	Stepan Comp.
Armosoft® Formulated ester-quaternary ammonium product [83]	Akzo Nobel
Atmos® 150 Glyceryl monostearate [89]	Witco Corp.
Baracarb® Ground marble [51]	Halliburton Energy Services, Inc.
Benol® White mineral oil [125]	Sonneborn Refined Products
Brij® (Series) Ethoxylated fatty alcohols [52]	ICI Surfactants
Caracarb® 5 Bridging agent [51]	Halliburton Energy Services, Inc.
Carnation® White mineral oil [125]	Sonneborn Refined Products
ClearFRAC™ Stimulating fluid [6, 10, 82, 83, 163]	Schlumberger Technology Corp.
Diamond FRAQ™ VES breaker [125]	Baker Hughes, Inc.
DiamondFRAQ™ VES system [163]	Baker Hughes, Inc.
Empol™ (Series) Oligomeric oleic acid [85, 86]	Henkel
Escaid® (Series) Mineral oils [125]	Crompton Corp.
Escaid® 110 Petroleum distillate [125]	Exxon Mobil
Flow-Back™ Xanthan gum, welan gum [22, 23]	Halliburton Energy Services, Inc.
Geltone® (Series) Organophilic clay [37]	Halliburton Energy Services, Inc.

Table 52 Tradenames in references—cont'd
Tradename
Description	Supplier
Gloria® High viscosity mineral oil [125]	Sonneborn Refined Products
Halad® (Series) Fluid loss control additive [52]	Halliburton Energy Services, Inc.
High dense® -3 Hematite-ore additive [52]	Halliburton Energy Services, Inc.
Hydrobrite® 200 White mineral oil [125]	Sonneborn, Inc.
Isopar® (Series) Isoparaffinic solvent [125]	Exxon Mobil
Jordapon® ACl Sodium cocoyl isothionate surfactant [163]	BASF AG
Jordapon® Cl Ammonium cocoyl isothionate surfactant [163]	BASF AG
Kaydol® oil Mineral oil [125]	Witco Corp.
Lutensol® AT (Series) Alkylpolyethylene glycol ethers (non-ionic surfactants) [52]	BASF AG
LVT® -200 Low viscosity oil for the formulation of invert emulsion drilling fluids [125]	Geo Drilling Fluids, Inc.
Micromax® Weighting agents [52]	Halliburton Energy Services, Inc.
Microsponge™ Porous solid substrate [10, 163]	Advanced Polymer Systems, Inc.
N-VIS® Viscosifier [51]	Halliburton Energy Services, Inc.
Poly-S.RTM Polymer encapsulation coating [10, 163]	Scotts Comp.
Rhodafac® LO-11A Phosphate ester [37]	Rhodia, Inc. Corp.
Semtol® Mineral oil [125]	Crompton Corp.
Silicalite® High surface area amorphous silica [52]	Halliburton Energy Services, Inc.
Span® 20 Sorbitan monolaurate [125]	Uniqema

Continued

Table 52 Tradenames in references—cont'd
Tradename

Description	Supplier
Span® 40 Sorbitan monopalmitate [125]	Uniqema
Span® 61 Sorbitan monostearate [125]	Uniqema
Span® 65 Sorbitan tristearate [125]	Uniqema
Span® 80 Sorbitan monooleate [125]	Uniqema
Span® 85 Sorbitan trioleate [125]	Uniqema
SurFRAQ™ VES Tallow amido propylamine oxide [125]	Baker Hughes, Inc.
Teflon® AF Amorphous copolymers of perfluoro-2,2-dimethyl-1,3-dioxole with tetrafluoroethylene [64]	DuPont
Teflon® Tetrafluoro polymer [1, 64]	DuPont
Tween® (Series) Ethoxylated fatty acid ester surfactants [125]	Uniqema
Tween® 20 Sorbitan monolaurate [125]	Uniqema
Tween® 21 Sorbitan monolaurate [125]	Uniqema
Tween® 40 Sorbitan monopalmitate [125]	Uniqema
Tween® 60 Sorbitan monostearate [125]	Uniqema
Tween® 61 Sorbitan monostearate [125]	Uniqema
Tween® 65 Sorbitan tristearate [125]	Uniqema
Tween® 81 Sorbitan monooleate [125]	Uniqema
Tween® 85 Sorbitan monooleate [125]	Uniqema
VES-STA 1 Gel stabilizer [125]	Baker Hughes, Inc.

Table 52 Tradenames in references—cont'd

Tradename Description	Supplier
	Halliburton Energy Services, Inc.
Well life® (Series) Fibers for cement improvement [52]	
Wellguard™ 7137 Interhalogen gel breaker [120, 160]	Albemarle Corp.
WG-3L VES-AROMOX® APA-T Viscoelastic surfactant [125]	Akzo Nobel
WS-44 Emulsifier [31]	Halliburton Energy Services, Inc.

REFERENCES

[1] Sierra L, Eoff LS. Method useful for controlling fluid loss in subterranean formations. US Patent 8181703, assigned to Halliburton Energy Services, Inc. (Duncan, OK); 2012. URL: http://www.freepatentsonline.com/8181703.html.

[2] Ebinger CD, Hunt E. Keys to good fracturing: Pt. 6: new fluids help increase effectiveness of hydraulic fracturing. Oil Gas J 1989;87(23):52-5.

[3] Ely JW. Fracturing fluids and additives. In: Doherty HL, editor. Recent advances in hydraulic fracturing (SPE monogr ser), vol. 12. Richardson (TX): SPE; 1989. ISBN 1-55563-020-0.

[4] Lemanczyk ZR. The use of polymers in well stimulation: performance, availability and economics. In: Proceedings volume. Plast rubber inst use of polymers in drilling & oilfield fluids conf. (London, England, 12/9/91); 1991.

[5] Li F, Dahanayake M, Colaco A. Multicomponent viscoelastic surfactant fluid and method of using as a fracturing fluid. US Patent 7772164, assigned to Rhodia, Inc. (Cranbury, NJ); 2010. URL: http://www.freepatentsonline.com/7772164.html.

[6] Couillet I, Hughes T. Aqueous fracturing fluid. US Patent 7427583, assigned to Schlumberger Technology Corporation (Ridgefield, CT); 2008. URL: http://www.freepatentsonline.com/7427583.html.

[7] Jones TGJ, Tustin GJ. Process of hydraulic fracturing using a viscoelastic wellbore fluid. US Patent 7655604, assigned to Schlumberger Technology Corporation (Ridgefield, CT); 2010. URL: http://www.freepatentsonline.com/7655604.html.

[8] McElfresh PM, Williams CF. Hydraulic fracturing using non-ionic surfactant gelling agent. US Patent 7216709, assigned to Akzo Nobel N.V. (Arnhem, NL); 2007. URL: http://www.freepatentsonline.com/7216709.html.

[9] Crews JB, Huang T, Gabrysch AD, Treadway JH, Willingham JR, Kelly PA, et al. Methods and compositions for fracturing subterranean formations. US Patent 7723272, assigned to Baker Hughes Incorporated (Houston, TX); 2010. URL: http://www.freepatentsonline.com/7723272.html.

[10] Crews JB. Bacteria-based and enzyme-based mechanisms and products for viscosity reduction breaking of viscoelastic fluids. US Patent 7052901, assigned to Baker Hughes Incorporated (Houston, TX); 2006. URL: http://www.freepatentsonline.com/7052901.html.

[11] Harms WM, Norman LR. Concentrated hydrophilic polymer suspensions. US Patent 4772646, assigned to Halliburton Company (Duncan, OK); 1988. URL: http://www.freepatentsonline.com/4772646.html.

[12] Bharat P. Well treating fluids and additives therefor. EP Patent 0372469 assigned to Phillips Petroleum Company; 1990. URL: https://www.google.at/patents/EP0372469A2?cl=en.

[13] Cooke Jr CE. Method and materials for hydraulic fracturing of wells using a liquid degradable thermoplastic polymer. US Patent 7569523; 2009. URL: http://www.freepatentsonline.com/7569523.html.

[14] Pelissier JJM, Biasini S. Biodegradable drilling mud (boue de forage biodegradable). FR Patent 2649988; 1991.

[15] Guo DR, Gao JP, Lu KH, Sun MB, Wang W. Study on the biodegradability of mud additives. Drill Fluid Completion Fluid 1996;13(1):10–12.

[16] Gregory G, Shuell D, Thompson Sr JE. Overview of contemporary LFC (liquid frac concentrate) fracture treatment systems and techniques. In: Proceedings volume. 91-01; 4th Cade/caodc spring drilling conf. (Calgary, Can, 4/10–12/91); 1991.

[17] Harms WM, Watts M, Venditto J, Chisholm P. Diesel-based HPG (hydroxypropyl guar) concentrate is product of evolution. Pet Eng Int 1988;60(4):51–4.

[18] Brannon HD. Fracturing fluid slurry concentrate and method of use. EP Patent 0280341 assigned to Pumptech N.V., Compagnie Des Services Dowell Schlumberger; 1988. URL: https://www.google.at/patents/EP0280341A1?cl=en.

[19] Burdick CL, Pullig JN. Sodium formate fluidized polymer suspensions process. US Patent 5228908, assigned to Aqualon Company (Wilmington, DE); 1993. URL: http://www.freepatentsonline.com/5228908.html.

[20] Blauer RE, Durborow CJ. Formation fracturing with stable foam. US Patent 3937283, assigned to The Dow Chemical Company (Midland, MI) Minerals Management, Inc. (Denver, CO); 1976. URL: http://www.freepatentsonline.com/3937283.html.

[21] Pakulski M, Hlidek BT. Slurried polymer foam system and method for the use thereof. US Patent 5360558, assigned to The Western Company of North America (Houston, TX); 1994. URL: http://www.freepatentsonline.com/5360558.html.

[22] Chatterji J, Crook R, King KL. Foamed fracturing fluids, additives and methods of fracturing subterranean zones. US Patent 6454008, assigned to Halliburton Energy Services, Inc. (Duncan, OK); 2002. URL: http://www.freepatentsonline.com/6454008.html.

[23] Chatterji J, Crook R, King KL. Foamed fracturing fluids, additives and methods of fracturing subterranean zones. US Patent 6734146, assigned to Halliburton Energy Services, Inc. (Duncan, OK); 2004. URL: http://www.freepatentsonline.com/6734146.html.

[24] Xiao B, Zhang S, Zhang J, Hou T, Guo T, Kaiyu L. Experimental investigation of a novel high temperature resistant and low friction fracturing fluid. Physicochem Probl Miner Process 2014;51(1):37–47. URL: http://www.minproc.pwr.wroc.pl/journal/pdf/ppmp51-1.37-47.pdf.

[25] Rostami A, Nguyen DT, Nasr-El-Din HA. Improving gas relative permeability in tight gas formations by using microemulsions. IPTC-17675-MS; International Petroleum Technology Conference, 19–22 January, Doha, Qatar; Richardson, TX: International Petroleum Technology Conference; 2014, URL: https://www.onepetro.org/conference-paper/IPTC-17675-MS. doi:http://dx.doi.org/10.2523/17675-MS.

[26] Berger PD, Berger CH. Environmental friendly fracturing and stimulation composition and method of using the same. US Patent 7998911, assigned to Oil Chem. Technologies (Sugar Land, TX); 2011. URL: http://www.freepatentsonline.com/7998911.html.

[27] Anonymous. Fracturing products and additives. World Oil 1999;220(8):135, 137, 139–45.

[28]   Harris P. Fracturing-fluid additives. J Pet Technol 1988;40(10). doi:10.2118/17112-PA.

[29]   Hrachovy MJ. Hydraulic fracturing technique employing in situ precipitation. WO Patent 9406998, assigned to Union Oil Co. California; 1994.

[30]   Hrachovy MJ. Hydraulic fracturing technique employing in situ precipitation. US Patent 5322121, assigned to Union Oil Company of California (Los Angeles, CA); 1994. URL: http://www.freepatentsonline.com/5322121.html.

[31]   Welton TD, Todd BL, McMechan D. Methods for effecting controlled break in pH dependent foamed fracturing fluid. US Patent 7662756, assigned to Halliburton Energy Services, Inc. (Duncan, OK); 2010. URL: http://www.freepatentsonline.com/7662756.html.

[32]   Lemanczyk ZR. The use of polymers in well stimulation: an overview of application, performance and economics. Oil Gas Europe Mag 1992;18(3):20–6.

[33]   Mondshine TC. Crosslinked fracturing fluids. WO Patent 8700236, assigned to Texas United Chemical Corp.; 1987.

[34]   Holtmyer MD, Hunt CV. Crosslinkable cellulose derivatives. EP Patent 0479606 assigned to Halliburton Company; 1995. URL: https://www.google.at/patents/EP0479606B1?cl=en.

[35]   Westland JA, Lenk DA, Penny GS. Rheological characteristics of reticulated bacterial cellulose as a performance additive to fracturing and drilling fluids. In: Proceedings volume. SPE oilfield chem. int. symp. (New Orleans, 3/2–5/93); 1993, p. 501–14.

[36]   Hodge RM. Particle transport fluids thickened with acetylate free xanthan heteropolysaccharide biopolymer plus guar gum. US Patent 5591699, assigned to E.I. du Pont de Nemours and Company (Wilmington, DE); 1997. URL: http://www.freepatentsonline.com/5591699.html.

[37]   Lawrence S, Warrender N. Crosslinking composition for fracturing fluids. US Patent 7749946, assigned to Sanjel Corporation (Calgary, Alberta, CA); 2010. URL: http://www.freepatentsonline.com/7749946.html.

[38]   Müller H, Herold CP, von Tapavicza S. Aqueous swellable compositions of guar gum and guar gum derivatives in oleophilic liquids and their use. US Patent 6180572, assigned to Henkel Kommanditgesellschaft auf Aktien (Düsseldorf, DE); 2001. URL: http://www.freepatentsonline.com/6180572.html.

[39]   Zeilinger SC, Mayerhofer MJ, Economides MJ. A comparison of the fluid-loss properties of borate-zirconate-crosslinked and noncrosslinked fracturing fluids. In: Proceedings volume. SPE east reg conf. (Lexington, KY, 10/23–25/91); 1991, p. 201–9.

[40]   Kelly PA, Gabrysch AD, Horner DN. Stabilizing crosslinked polymer guars and modified guar derivatives. US Patent 7195065, assigned to Baker Hughes Incorporated (Houston, TX); 2007. URL: http://www.freepatentsonline.com/7195065.html.

[41]   Yeh MH. Anionic sulfonated thickening compositions. EP Patent 0632057 assigned to Rhone-Poulenc Specialty Chemicals Co.; 1995. URL: https://www.google.at/patents/EP0632057A1?cl=en.

[42]   Doherty DH, Ferber DM, Marrelli JD, Vanderslice RW, Hassler RA. Genetic control of acetylation and pyruvylation of xanthan based polysaccharide polymers. WO Patent 9219753, assigned to Getty Scientific Dev Co.; 1992.

[43]   Penny GS, Stephens RS, Winslow AR. Method of supporting fractures in geologic formations and hydraulic fluid composition for same. US Patent 5009797, assigned to Weyerhaeuser Company (Tacoma, WA); 1991. URL: http://www.freepatentsonline.com/5009797.html.

[44]   Aggour TM, Economides MJ. Impact of fluid selection on high-permeability fracturing. In: Proceedings volume; vol. 2. SPE Europe petrol. conf. (Milan, Italy, 10/22–24/96); 1996, p. 281–7.

[45] Johnson M. Fluid systems for controlling fluid losses during hydrocarbon recovery operations. EP Patent 0691454 assigned to Baker Hughes Incorporated; 1999. URL: https://www.google.at/patents/EP0691454B1?cl=en.

[46] Johnson MH, Smejkal KD. Fluid system for controlling fluid losses during hydrocarbon recovery operations. US Patent 5228524, assigned to Baker Hughes Incorporated (Houston, TX); 1993. URL: http://www.freepatentsonline.com/5228524. html.

[47] Elbel JL, Navarrete RC, Poe Jr BD. Production effects of fluid loss in fracturing high-permeability formations. In: Proceedings volume. SPE Europe formation damage contr. conf. (The Hague, The Netherlands, 5/15–16/95); 1995, p. 201–11.

[48] Navarrete RC, Brown JE, Marcinew RP. Application of new bridging technology and particulate chemistry for fluid-loss control during fracturing highly permeable formations. In: Proceedings volume; vol. 2. SPE Europe petrol. conf. (Milan, Italy, 10/22–24/96); 1996, p. 321–5.

[49] Navarrete RC, Mitchell JP. Fluid-loss control for high-permeability rocks in hydraulic fracturing under realistic shear conditions. In: Proceedings volume. SPE prod. oper. symp. (Oklahoma City, 4/2–4/95); 1995, p. 579–91.

[50] Cawiezel KE, Navarrete RC, Constien VG. Fluid loss control. US Patent 5948733, assigned to Dowell Schlumberger Incorporated (Sugar Land, TX); 1999. URL: http://www.freepatentsonline.com/5948733.html.

[51] Ezell RG, Wu JJ. Methods of using fluid loss additives comprising micro gels. US Patent 8697609, assigned to Halliburton Energy Services, Inc. (Houston, TX); 2014. URL: http://www.freepatentsonline.com/8697609.html.

[52] Tarafdar A, Ak R, Patil RC, Wagle V. Water based fluid loss additive containing an amphiphilic dispersant for use in a well. US Patent 8741817, assigned to Halliburton Energy Services, Inc. (Houston, TX); 2014. URL: http://www.freepatentsonline. com/8741817.html.

[53] Williamson CD, Allenson SJ. A new nondamaging particulate fluid-loss additive. In: Proceedings volume. SPE oilfield chem. int. symp. (Houston, 2/8–10/89); 1989, p. 147–58.

[54] Williamson CD, Allenson SJ, Gabel RK, Huddleston DA. Enzymatically degradable fluid loss additive. US Patent 5032297, assigned to Nalco Chemical Company (Naperville, IL); 1991. URL: http://www.freepatentsonline.com/5032297.html.

[55] Hofstaetter H. Permeable fracturing material. US Patent Application 20130206407, assigned to Montanuniversitaet Leoben, Leoben (AT); 2013. URL: http://www. freepatentsonline.com/20130206407.html.

[56] Dobson JW, Mondshine KB. Method of reducing fluid loss of well drilling and servicing fluids. EP Patent 0758011 assigned to Texas United Chemical Company, LLC.; 2001. URL: https://www.google.at/patents/EP0758011B1?cl=en.

[57] Williamson CD, Allenson SJ, Gabel RK. Additive and method for temporarily reducing permeability of subterranean formations. US Patent 4997581, assigned to Nalco Chemical Company (Naperville, IL); 1991. URL: http://www. freepatentsonline.com/4997581.html.

[58] Nimerick KH. Metal ion crosslinked fracturing fluid and method. US Patent 6177385, assigned to Schlumberger Technology Corporation (Sugar Land, TX); 2001. URL: http://www.freepatentsonline.com/6177385.html.

[59] Himes RE, Vinson EF, Simon DE. Clay stabilization in low-permeability formations. In: Proceedings volume. SPE prod. oper. symp. (Oklahoma City, 3/12–14/89); 1989, p. 507–16.

[60] Yeager RR, Bailey DE. Diesel-based gel concentrate improves rocky mountain region fracture treatments. In: Proceedings volume. SPE Rocky Mountain reg mtg. (Casper, Wyo, 5/11–13/88); 1988, p. 493–7.

[61] Thomas TR, Smith KW. Method of maintaining subterranean formation permeability and inhibiting clay swelling. US Patent 5211239, assigned to Clearwater, Inc. (Pittsburgh, PA); 1993. URL: http://www.freepatentsonline.com/5211239. html.

[62] Himes RE. Method for clay stabilization with quaternary amines. US Patent 5097904, assigned to Halliburton Company (Duncan, OK); 1992. URL: http://www.freepatentsonline.com/5097904.html.

[63] Himes RE, Vinson EF. Fluid additive and method for treatment of subterranean formations. US Patent 4842073, assigned to Halliburton Services (Duncan, OK); 1989. URL: http://www.freepatentsonline.com/4842073.html.

[64] Schield JA, Naiman MI, Scherubel GA. Polyimide quaternary salts as clay stabilization agents. US Patent 5160642, assigned to Petrolite Corporation (St. Louis, MO); 1992. URL: http://www.freepatentsonline.com/5160642.html.

[65] Aften CW, Gabel RK. Clay stabilizing method for oil and gas well treatment. US Patent 5099923, assigned to Nalco Chemical Company (Naperville, IL); 1992. URL: http://www.freepatentsonline.com/5099923.html.

[66] Hall BE, Szememyei CA. Fluid additive and method for treatment of subterranean formations. US Patent 5089151, assigned to The Western Company of North America (Houston, TX); 1992. URL: http://www.freepatentsonline.com/5089151. html.

[67] Himes RE, Parker MA, Schmelzl EG. Environmentally safe temporary clay stabilizer for use in well service fluids. In: Proceedings volume; vol. 3. Cim. petrol. soc/SPE int. tech. mtg. (Calgary, Can, 6/10–13/90); 1990.

[68] Himes RE, Vinson EF. Environmentally safe salt replacement for fracturing fluids. In: Proceedings volume. SPE east reg conf. (Lexington, KY, 10/23–25/91); 1991, p. 237–48.

[69] Poelker DJ, McMahon J, Schield JA. Polyamine salts as clay stabilizing agents. US Patent 7601675, assigned to Baker Hughes Incorporated (Houston, TX); 2009. URL: http://www.freepatentsonline.com/7601675.html.

[70] Kanda S, Yanagita M, Sekimoto Y. Stabilized fracturing fluid and method of stabilizing fracturing fluid. US Patent 4681690, assigned to Nitto Chemical Industry Co. , Ltd. (Tokyo, JP); 1987. URL: http://www.freepatentsonline.com/4681690. html.

[71] Kanda S, Kawamura Z. Stabilization of xanthan gum in aqueous solution. GB Patent 2192402, assigned to Nitto Chemical Industry Co. Ltd.; 1988.

[72] Kanda S, Kawamura Z. Stabilization of xanthan gum in aqueous solution. US Patent 4810786, assigned to Nitto Chemical Industry Co. , Ltd. (Tokyo, JP); 1989. URL: http://www.freepatentsonline.com/4810786.html.

[73] Kanda S, Yanagita M, Sekimoto Y. Stabilized fracturing fluid and method of stabilizing fracturing fluid. US Patent 4721577, assigned to Nitto Chemical Industry Co. , Ltd. (Tokyo, JP); 1988. URL: http://www.freepatentsonline.com/4721577. html.

[74] Penny GS. Method of increasing hydrocarbon production from subterranean formations. US Patent 4702849, assigned to Halliburton Company (Duncan, OK); 1987. URL: http://www.freepatentsonline.com/4702849.html.

[75] Penny GS. Method of increasing hydrocarbon productions from subterranean formations. EP Patent 0234910 assigned to Halliburton Company; 1987. URL: https://www.google.at/patents/EP0234910A2?cl=en.

[76] Penny GS, Briscoe JE. Method of increasing hydrocarbon production by remedial well treatment. CA Patent 1216416 assigned to Glenn S. Penny, Halliburton Company, James E. Briscoe; 1987. URL: https://www.google.at/patents/CA1216416A1?cl=en.

[77] Norman WD, Jasinski RJ, Nelson EB. Hydraulic fracturing process and compositions. US Patent 5551516, assigned to Dowell, a division of Schlumberger Technology Corporation; 1996. URL: http://www.freepatentsonline.com/5551516.html.

[78] Hughes TL, Jones TGJ, Tustin GJ. Viscoelastic surfactant based gelling composition for wellbore service fluids. US Patent 6232274, assigned to Schlumberger Technology Corporation (Sugar Land, TX); 2001. URL: http://www.freepatentsonline. com/6232274.html.

[79] Jones TGJ, Tustin GJ. Gelling composition for wellbore service fluids. US Patent 6194356, assigned to Schlumberger Technology Corporation (Sugar Land, TX); 2001. URL: http://www.freepatentsonline.com/6194356.html.

[80] Davies SN, Jones TGJ, Olthoff S, Tustin GJ. Method for water control. US Patent 6920928, assigned to Schlumberger Technology Corporation (Sugar Land, TX); 2005. URL: http://www.freepatentsonline.com/6920928.html.

[81] Jones TGJ, Tustin GJ, Fletcher P, Lee JCW. Scale dissolver fluid. US Patent 7156177, assigned to Schlumberger Technology Corporation (Ridgefield, CT); 2007. URL: http://www.freepatentsonline.com/7156177.html.

[82] Brown JE, Card RJ, Nelson EB. Methods and compositions for testing subterranean formations. US Patent 5964295, assigned to Schlumberger Technology Corporation, Dowell division (Sugar Land, TX); 1999. URL: http://www.freepatentsonline.com/ 5964295.html.

[83] Zhou J, Hughes T. Aqueous viscoelastic fluid. US Patent 7036585, assigned to Schlumberger Technology Corporation (Ridgefield, CT); 2006. URL: http://www. freepatentsonline.com/7036585.html.

[84] Hartshorne RS, Hughes TL, Jones TGJ, Tustin GJ. Anionic viscoelastic surfactant. US Patent Application 20050124525, assigned to Schlumberger Technology Corporation (Ridgefield, CT); 2005. URL: http://www.freepatentsonline.com/ 20050124525.html.

[85] Jones TGJ, Tustin GJ, Fletcher P, Lee JCW. Scale dissolver fluid. US Patent 7343978, assigned to Schlumberger Technology Corporation (Ridgefield, CT); 2008. URL: http://www.freepatentsonline.com/7343978.html.

[86] Jones TGJ, Tustin GJ. Powder composition. US Patent 7858562, assigned to Schlumberger Technology Corporation (Ridgefield, CT); 2010. URL: http://www. freepatentsonline.com/7858562.html.

[87] Koganei R. On fatty acids obtained from cephalin. Compounds of β-aminoethyl alcohol with saturated and unsaturated fatty acids. J Biochem 1923;3(1):15–26.

[88] Sinha KR, Caldwell BE. Glass coating composition and method. US Patent 4517243, assigned to Wheaton Industries (Millville, NJ); 1985. URL: http://www. freepatentsonline.com/4517243.html.

[89] Knaus DA. Stability control agent composition for polyolefin foam. US Patent 5874024; 1999. URL: http://www.freepatentsonline.com/5874024.html.

[90] Tsuji K, Yamamoto M, Kawamoto K, Tachibana H. Method for treating an allergic or inflammatory disease. US Patent 6491943, assigned to National Agricultural Research Organization (Tsukuba, JP); 2002. URL: http://www.freepatentsonline. com/6491943.html.

[91] Tsuji K, Yamamoto M, Kawamoto K, Tachibana H. Cosmetics, foods and beverages supplemented with purified strictinin. US Patent 6638524, assigned to National Agricultural Research Organization (Tsukuba, JP) Bio-Oriented Technology Research Advancement Institution (Omiya, JP); 2003. URL: http://www. freepatentsonline.com/6638524.html.

[92] Takahata K, Matsui Y. Antialopecia agent. US Patent 6713093, assigned to Suntory Limited (Osaka, JP); 2004. URL: http://www.freepatentsonline.com/6713093. html.

[93] Tsuji K, Yamamoto M, Kawamoto K, Tachibana H. Method for treating an allergic or inflammatory disease. US Patent 6899893, assigned to National Agriculture Research Organization (Tsukuba, JP), Bio-Oriented Technology Research Advancement Institution (Omiya, JP); 2005. URL: http://www.freepatentsonline.com/6899893.html.

[94] Farmer RF, Doyle AK, Vale GDC, Gadberry JF, Hoey MD, Dobson RE. Method for controlling the rheology of an aqueous fluid and gelling agent therefor. US Patent 6239183, assigned to Akzo Nobel N.V. (Arnhem, NL); 2001. URL: http://www.freepatentsonline.com/6239183.html.

[95] Hoey MD, Franklin R, Lucas DM, Dery M, Dobson RE, Engel M, et al. Viscoelastic surfactants and compositions containing same. US Patent 6506710, assigned to Akzo Nobel N.V. (Arnhem, NL); 2003. URL: http://www.freepatentsonline.com/6506710.html.

[96] Hartshorne RS, Hughes TL, Jones TGJ, Tustin GJ, Westwood JF. Wellbore treatment fluid. US Patent 8252730, assigned to Schlumberger Technology Corporation (Sugar Land, TX); 2012. URL: http://www.freepatentsonline.com/8252730.html.

[97] Gadberry JF, Hoey MD, Franklin R, del Carmen Vale G, Mozayeni F. Surfactants for hydraulic fracturing compositions. US Patent 5979555, assigned to Akzo Nobel N.V. (Arnhem, NL); 1999. URL: http://www.freepatentsonline.com/5979555.html.

[98] Barkat O. Rheology of flowing, reacting systems: the crosslinking reaction of hydroxypropyl guar with titanium chelates [Ph.D. thesis]. Tulsa Univ; 1987.

[99] Le HV, Wood WR. Method for increasing the stability of water based fracturing fluids. US Patent 5226481, assigned to BJ Services Company (Houston, TX); 1993. URL: http://www.freepatentsonline.com/5226481.html.

[100] Ainley BR, Nimerick KH, Card RJ. High-temperature, borate-crosslinked fracturing fluids: a comparison of delay methodology. In: Proceedings volume. SPE prod. oper. symp. (Oklahoma City, 3/21–23/93); 1993, p. 517–20.

[101] Cawiezel KE, Elbel JL. A new system for controlling the crosslinking rate of borate fracturing fluids. In: Proceedings volume. 60th Annu. SPE Calif reg mtg. (Ventura, Calif, 4/4–6/90); 1990, p. 547–52.

[102] Harris PC, Norman LR, Hollenbeak KH. Borate crosslinked fracturing fluids. EP Patent 0594363 assigned to Halliburton Company; 1994. URL: https://www.google.at/patents/EP0594363A1?cl=en.

[103] Harris PC, Heath SJ. Delayed release borate crosslinking agent. US Patent 5372732, assigned to Halliburton Company (Duncan, OK); 1994. URL: http://www.freepatentsonline.com/5372732.html.

[104] Sanner T, Kightlinger AP, Davis JR. Borate-starch compositions for use in oil field and other industrial applications. US Patent 5559082, assigned to Grain Processing Corporation (Muscatine, IA); 1996. URL: http://www.freepatentsonline.com/5559082.html.

[105] Sun H, Qu Q. High-efficiency boron crosslinkers for low-polymer fracturing fluids. SPE Int Symp Oilfield Chem 2011. doi:10.2118/140817-ms.

[106] Legemah M, Sun H, Guerin M, Qu Q. Novel high-efficiency boron crosslinkers for low-polymer-loading fracturing fluids. SPE J 2014;19(4):737–43. doi:10.2118/164118-pa.

[107] Dawson JC. Method and composition for delaying the gellation of borated galactomannans. US Patent 5082579, assigned to BJ Services Company (Houston, TX); 1992. URL: http://www.freepatentsonline.com/5082579.html.

[108] Dawson JC. Method for delaying the gellation of borated galactomannans with a delay additive such as glyoxal. US Patent 5160643, assigned to BJ Services Company (Houston, TX); 1992. URL: http://www.freepatentsonline.com/5160643.html.

[109] Dawson JC. Method and composition for delaying the gellation of borated gallactomannans. CA Patent 2037974 assigned to Jeffrey C. Dawson, Bj Services Company; 2001. URL: https://www.google.at/patents/CA2037974C?cl=en.

[110] Nimerick KH, Crown CW, McConnell SB, Ainley B. Method of using borate crosslinked fracturing fluid having increased temperature range. US Patent 5259455; 1993. URL: http://www.freepatentsonline.com/5259455.html.

[111] Ainley BR, McConnell SB. Method of fracturing a subterranean formation. EP Patent 0528461 assigned to Compagnie Des Services Dowell Schlumberger S.A., Pumptech N.V.; 2002. URL: https://www.google.at/patents/EP0528461B2?cl=en.

[112] Putzig DE, Smeltz KC. Organic titanium compositions useful as cross-linkers. EP Patent 195531; 1986.

[113] Putzig DE. Zirconium chelates and their use for cross-linking. EP Patent 0278684 assigned to E.I. du Pont de Nemours and Company; 1992. URL: https://www.google.at/patents/EP0278684B1?cl=en.

[114] Ridland J, Brown DA. Organo-metallic compounds. CA Patent 2002792 assigned to John Ridland, David Alexander Brown, Tioxide Group Plc, Tioxide Group Limited, Tioxide Group, Acma Limited; 1997. URL: https://www.google.at/patents/CA2002792C?cl=en.

[115] Dawson JC, Le HV. Gelation additive for hydraulic fracturing fluids. US Patent 5798320, assigned to BJ Services Company (Houston, TX); 1998. URL: http://www.freepatentsonline.com/5798320.html.

[116] Dawson JC, Le HV. Gelation additive for hydraulic fracturing fluids. US Patent 5773638, assigned to BJ Services Company (Houston, TX); 1998. URL: http://www.freepatentsonline.com/5773638.html.

[117] Sharif S. Process for preparation and composition of stable aqueous solutions of boron zirconium chelates for high temperature frac fluids. US Patent 5217632, assigned to Zirconium Technology Corporation (Midland, TX); 1993. URL: http://www.freepatentsonline.com/5217632.html.

[118] Sharif S. Process for preparation of stable aqueous solutions of zirconium chelates. US Patent 5466846, assigned to Benchmark Research and Technology, Inc. (San Antonio, TX); 1995. URL: http://www.freepatentsonline.com/5466846.html.

[119] Putzig DE. Process to prepare borozirconate solution and use as cross-linker in hydraulic fracturing fluids. US Patent 7683011; 2010. URL: http://www.freepatentsonline.com/7683011.html.

[120] Carpenter JF. Bromine-based sulfamate stabilized breaker composition and process. US Patent 7576041, assigned to Albemarle Corporation (Baton Rouge, LA); 2009. URL: http://www.freepatentsonline.com/7576041.html.

[121] Craig D, Holditch SA. The degradation of hydroxypropyl guar fracturing fluids by enzyme, oxidative, and catalyzed oxidative breakers: Pt. 2: crosslinked hydroxypropyl guar gels: topical report (January 1992–April 1992). Gas Res Inst Rep GRI-93/04192; Gas Res Inst; 1993.

[122] Craig D, Holditch SA. The degradation of hydroxypropyl guar fracturing fluids by enzyme, oxidative, and catalyzed oxidative breakers: Pt. 1: linear hydroxypropyl guar solutions: topical report (February 1991–December 1991). Gas Res Inst Rep GRI-93/04191; Gas Res Inst; 1993.

[123] Craig DP. The degradation of hydroxypropyl guar fracturing fluids by enzyme, oxidative, and catalyzed oxidative breakers [Ph.D. thesis]. Texas A & M Univ; 1991.

[124] Bielewicz VD, Kraj L. Laboratory data on the effectivity of chemical breakers in mud and filtercake (Untersuchungen zur Effektivität von Degradationsmitteln in Spülungen). Erdöl Erdgas Kohle 1998;114(2):76–9.

[125] Huang T, Crews JB. Dual-functional breaker for hybrid fluids of viscoelastic surfactant and polymer. US Patent 8383557, assigned to Baker Hughes Incorporated (Houston, TX); 2013. URL: http://www.freepatentsonline.com/8383557.html.

[126]   Williams MM, Phelps MA, Zody GM. Reduction of viscosity of aqueous fluids. EP Patent 0222615 assigned to Hi-Tek Polymers, Inc.; 1990. URL: https://www. google.at/patents/EP0222615B1?cl=en.

[127]   Langemeier PW, Phelps MA, Morgan ME. Method for reducing the viscosity of aqueous fluids. EP Patent 0330489 assigned to Stein, Hall & Co., Inc.; 1989. URL: https://www.google.at/patents/EP0330489A2?cl=en.

[128]   Walker ML, Shuchart CE. Method for breaking stabilized viscosified fluids. US Patent 5413178, assigned to Halliburton Company (Duncan, OK); 1995. URL: http://www.freepatentsonline.com/5413178.html.

[129]   Mondshine TC. Process for decomposing polysaccharides in alkaline aqueous systems. EP Patent 0559418 assigned to Texas United Chemical Company, LLC.; 1997. URL: https://www.google.at/patents/EP0559418B1?cl=en.

[130]   Laramay SB, Powell RJ, Pelley SD. Perphosphate viscosity breakers in well fracture fluids. US Patent 5386874, assigned to Halliburton Company (Duncan, OK); 1995. URL: http://www.freepatentsonline.com/5386874.html.

[131]   Harms WM. Catalyst for breaker system for high viscosity fluids. US Patent 5143157, assigned to Halliburton Company (Duncan, OK); 1992. URL: http:// www.freepatentsonline.com/5143157.html.

[132]   Dawson JC, Le HV. Controlled degradation of polymer based aqueous gels. US Patent 5447199, assigned to BJ Services Company (Houston, TX); 1995. URL: http://www.freepatentsonline.com/5447199.html.

[133]   Mccabe MA, Shuchart CE, Slabaugh BF, Terracina JM. Method of treating subterranean formation. EP Patent 0916806 assigned to Halliburton Energy Services, Inc.; 1999. URL: https://www.google.at/patents/EP0916806A2?cl=en.

[134]   Noran L, Vitthal S, Terracina J. New breaker technology for fracturing high-permeability formations. In: Proceedings volume. SPE Europe formation damage contr. conf. (The Hague, The Netherlands, 5/15–16/95); 1995, p. 187–99.

[135]   Cantu LA, Boyd PA. Laboratory and field evaluation of a combined fluid-loss control additive and gel breaker for fracturing fluids. In: Proceedings volume. SPE oilfield chem. int. symp. (Houston, 2/8–10/89); 1989, p. 7–16.

[136]   Cantu LA, McBride EFMO. Formation fracturing process. EP Patent 0401431 assigned to Conoco, Inc., E.I. du Pont de Nemours and Company; 1990. URL: https://www.google.at/patents/EP0401431A1?cl=en.

[137]   Cantu LA, McBride EF, Osborne M. Well treatment process. EP Patent 0404489 assigned to Conoco, Inc., E.I. du Pont de Nemours and Company; 1995. URL: https://www.google.at/patents/EP0404489B1?cl=en.

[138]   Cantu LA, Mcbride EF, Osborne MW. Formation fracturing process. CA Patent 1319819 assigned to Lisa A. Cantu, Edward F. Mcbride, Marion W. Osborne, Conoco, Inc., E.I. du Pont de Nemours and Company; 1993. URL: https://www. google.at/patents/CA1319819C?cl=en.

[139]   Brannon HD, Tjon-Joe-Pin RM. Biotechnological breakthrough improves performance of moderate to high-temperature fracturing applications. In: Proceedings volume; vol. 1. 69th Annu. SPE tech. conf. (New Orleans, 9/25–28/94); 1994, p. 515–30.

[140]   Slodki ME, Cadmus MC. High-temperature, salt-tolerant enzymic breaker of xanthan gum viscosity. In: Donaldson EC, editor. Microbial enhancement of oil recovery: recent advances: proceedings of the 1990 international conference on microbial enhancement of oil recovery; vol. 31 of Developments in petroleum science. Elsevier Science Ltd. ISBN 0-444-88633-8; 1991, p. 247–55.

[141]   Craig D, Holditch SA, Howard B. The degradation of hydroxypropyl guar fracturing fluids by enzyme, oxidative, and catalyzed oxidative breakers. In: Proceedings volume. 39th Annu. southwestern petrol. Short Course Ass. Inc. et al mtg. (Lubbock, TX, 4/22–23/92); 1992, p. 1–19.

[142]   Ahlgren JA. Enzymatic hydrolysis of xanthan gum at elevated temperatures and salt concentrations. In: Proceedings volume. 6th Inst. gas technol. gas, oil, & environ. biotechnol. int. symp. (Colorado Springs, CO, 11/29/93–12/1/93); 1993.

[143]   Fodge DW, Anderson DM, Pettey TM. Hemicellulase active at extremes of pH and temperature and utilizing the enzyme in oil wells. US Patent 5551515, assigned to Chemgen Corporation (Gaithersburg, MD); 1996. URL: http://www.freepatentsonline.com/5551515.html.

[144]   Harris RE, Hodgson RJ. Delayed acid for gel breaking. US Patent 5813466, assigned to Cleansorb Limited (GB); 1998. URL: http://www.freepatentsonline.com/5813466.html.

[145]   Prasek BB. Interactions between fracturing fluid additives and currently used enzyme breakers. In: Proceedings volume. 43rd Annu. southwestern petrol. Short Course Ass. Inc. et al mtg. (Lubbock, Texas, 4/17–18/96); 1996, p. 265–79.

[146]   Muir DJ, Irwin MJ. Encapsulated breakers, compositions and methods of use. WO Patent 9961747, assigned to 3M Innovative Propertie C.; 1999.

[147]   Gulbis J, King MT, Hawkins GW, Brannon HD. Encapsulated breaker for aqueous polymeric fluids. In: Proceedings volume. 9th SPE Formation Damage Contr. Symp. (Lafayette, LA, 2/22–23/90); 1990, p. 245–54.

[148]   Gulbis J, King MT, Hawkins GW, Brannon HD. Encapsulated breaker for aqueous polymeric fluids. SPE Prod Eng 1992;7(1):9–14.

[149]   Gulbis J, Williamson TDA, King MT, Constien VG. Method of controlling release of encapsulated breakers. EP Patent 0404211 assigned to Pumptech N.V., Compagnie Des Services Dowell Schlumberger; 1990. URL: https://www.google.at/patents/EP0404211A1?cl=en.

[150]   King MT, Gulbis J, Hawkins GW, Brannon HD. Encapsulated breaker for aqueous polymeric fluids. In: Proceedings volume; vol. 2. Cim. petrol. soc/SPE int. tech. mtg. (Calgary, Can, 6/10–13/90); 1990.

[151]   Gupta DVS, Prasek BB. Method for fracturing subterranean formations using controlled release breakers and compositions useful therein. US Patent 5437331, assigned to The Western Company of North America (Houston, TX); 1995. URL: http://www.freepatentsonline.com/5437331.html.

[152]   Boles JL, Metcalf AS, Dawson JC. Coated breaker for crosslinked acid. US Patent 5497830, assigned to BJ Services Company (Houston, TX); 1996. URL: http://www.freepatentsonline.com/5497830.html.

[153]   Satyanarayana Gupta DV, Cooney A. Encapsulations for treating subterranean formations and methods for the use thereof. WO Patent 9210640, assigned to Western Co. North America; 1992.

[154]   Manalastas PV, Drake EN, Kresge EN, Thaler WA, McDougall LA, Newlove JC, et al. Breaker chemical encapsulated with a crosslinked elastomer coating. US Patent 5110486, assigned to Exxon Research and Engineering Company (Florham Park, NJ); 1992. URL: http://www.freepatentsonline.com/5110486.html.

[155]   Hunt CV, Powell RJ, Carter ML, Pelley SD, Norman LR. Encapsulated enzyme breaker and method for use in treating subterranean formations. US Patent 5604186, assigned to Halliburton Company (Duncan, OK); 1997. URL: http://www.freepatentsonline.com/5604186.html.

[156]   Norman LR, Laramay SB. Encapsulated breakers and method for use in treating subterranean formations. US Patent 5373901, assigned to Halliburton Company (Duncan, OK); 1994. URL: http://www.freepatentsonline.com/5373901.html.

[157]   Norman LR, Turton R, Bhatia AL. Breaking fracturing fluid in subterranean formation. EP Patent 1152121 assigned to Halliburton Energy Services, Inc.; 2010. URL: https://www.google.at/patents/EP1152121B1?cl=en.

[158] Swarup V, Peiffer DG, Gorbaty ML. Encapsulated breaker chemical. US Patent 5580844, assigned to Exxon Research and Engineering Company (Florham Park, NJ); 1996. URL: http://www.freepatentsonline.com/5580844.html.

[159] Hawkins GW. Molecular weight reduction and physical consequences of chemical degradation of hydroxypropylguar in aqueous brine solutions. In: Proceedings 192nd ACS nat mtg; vol. 55. Am Chem Soc polymeric, mater sci eng div tech program (Anaheim, Calif, 9/7–12/86). ISBN 0-8412-0985-5; 1986, p. 588–93.

[160] Carpenter JF. Breaker composition and process. US Patent 7223719, assigned to Albemarle Corporation (Richmond, VA); 2007. URL: http://www.freepatentsonline.com/7223719.html.

[161] Crews JB. Aminocarboxylic acid breaker compositions for fracturing fluids. US Patent 7208529, assigned to Baker Hughes Incorporated (Houston, TX); 2007. URL: http://www.freepatentsonline.com/7208529.html.

[162] Crews JB. Polyols for breaking of fracturing fluid. US Patent 7160842, assigned to Baker Hughes Incorporated (Houston, TX); 2007. URL: http://www.freepatentsonline.com/7160842.html.

[163] Crews JB. Saponified fatty acids as breakers for viscoelastic surfactant-gelled fluids. US Patent 7728044, assigned to Baker Hughes Incorporated (Houston, TX); 2010. URL: http://www.freepatentsonline.com/7728044.html.

[164] McDougall LA, Malekahmadi F, Williams DA. Method of fracturing formations. EP Patent 0540204 assigned to Exxon Chemical Patents, Inc.; 1993. URL: https://www.google.at/patents/EP0540204A2?cl=en.

[165] Mitchell TO, Card RJ, Gomtsyan A. Cleanup additive. US Patent 6242390, assigned to Schlumberger Technology Corporation (Sugar Land, TX); 2001. URL: http://www.freepatentsonline.com/6242390.html.

[166] Acker DB, Malekahmadi F. Delayed release breakers in gelled hydrocarbons. US Patent 6187720; 2001. URL: http://www.freepatentsonline.com/6187720.html.

[167] Watkins DR, Clemens JJ, Smith JC, Sharma SN, Edwards HG. Use of scale inhibitors in hydraulic fracture fluids to prevent scale build-up. US Patent 5224543, assigned to Union Oil Company of California (Los Angeles, CA); 1993. URL: http://www.freepatentsonline.com/5224543.html.

[168] Barthorpe RT. The impairment of scale inhibitor function by commonly used organic anions. In: Proceedings volume. SPE oilfield chem. int. symp. (New Orleans, 3/2–5/93); 1993, p. 69–76.

[169] Powell PJ, Gdanski RD, McCabe MA, Buster DC. Controlled-release scale inhibitor for use in fracturing treatments. In: Proceedings volume. SPE oilfield chem. int. symp. (San Antonio, 2/14–17/95); 1995, p. 571–9.

[170] Powell RJ, Fischer AR, Gdanski RD, McCabe MA, Pelley SD. Encapsulated scale inhibitor for use in fracturing treatments. In: Proceedings volume. Annu. SPE tech. conf. (Dallas, 10/22–25/95); 1995, p. 557–63.

[171] Powell RJ, Fischer AR, Gdanski RD, McCabe MA, Pelley SD. Encapsulated scale inhibitor for use in fracturing treatments. In: Proceedings volume. SPE Permian basin oil & gas recovery conf. (Midland, TX, 3/27–29/96); 1996, p. 107–13.

[172] Wang W, Kan AT, Zhang F, Yan C, Tomson M. Measurement and prediction of thermal degradation of scale inhibitors. SPE J 2014. doi:10.2118/164047-pa.

[173] Graue A, Moe RW, Baldwin BA, Needham R. Water mixing during waterflood oil recovery: the effect of initial water saturation. SPE J 2012;17(1):43–52. URL: http://www.onepetro.org/mslib/app/Preview.do?paperNumber=SPE-149577-PA&societyCode=SPE.

[174] Korsbech U, Aage H, Hedegaard K, Andersen B, Springer N. Measuring and modeling the displacement of connate water in chalk core plugs during water injection. SPE Reserv Eval Eng 2006;9(3):259–65. doi:10.2118/78059-PA.

[175]   Guo-Qing T, Abbas F. Effect of viscous forces and initial water saturation on water injection in water-wet and mixed-wet fractured porous media. Soc Pet Eng. ISBN 9781555633486; 2000. doi:10.2118/59291-MS.

[176]   Zhou D, Jiang T, Feng J, Bian W, Liu Y, Zhao J. Research of water flooded performance and pattern in horizontal well with bottom-water drive reservoir. Soc Pet Eng. ISBN 9781613991114; 2004. doi:10.2118/2004-093.

[177]   Lord P, Weston M, Fontenelle L, Haggstrom J. Recycling water: case studies in designing fracturing fluids using flowback, produced, and nontraditional water sources. Soc Pet Eng. ISBN 9781613992722; 2013. doi:10.2118/165641-MS.

[178]   Nie R, Jia Y, Shen N, Qin X, Luo X, Zhang W, et al. A new method to discriminate effectively the influence of water injection and natural water influx upon reservoir pressure system: a case history from Cainan oilfield. Soc Pet Eng. ISBN 9781555632748; 2010. doi:10.2118/127282-MS.

[179]   Jin L, Wojtanowicz A, Hughes R. An analytical model for water coning control installation in reservoir with bottom water. Soc Pet Eng. ISBN 9781613991169; 2009. doi:10.2118/2009-098.

[180]   Eoff LS, Dalrymple ED, Reddy BR. Methods and compositions for the diversion of aqueous injection fluids in injection operations. US Patent 7563750, assigned to Halliburton Energy Services, Inc. (Duncan, OK); 2009. URL: http://www.freepatentsonline.com/7563750.html.

[181]   Eoff LS, Reddy BR, Dalrymple ED. Methods of reducing subterranean formation water permeability. US Patent 6476169, assigned to Halliburton Energy Services, Inc. (Duncan, OK); 2002. URL: http://www.freepatentsonline.com/6476169.html.

[182]   Rabie AI, Shedd DC, Nasr-El-Din HA. Measuring the reaction rate of lactic acid with calcite and dolomite by use of the rotating-disk apparatus. SPE J 2014. doi:10.2118/140167-pa.

CHAPTER IV

Other Water-Based Uses

1. COMPLETION AND WORKOVER OPERATIONS

Completion fluids are fluids used during drilling and during the steps of completion, or recompletion, of the well. A completion operation can include [1, 2]:

- perforating the casing,
- setting the tubing and pump, and
- cementing the casing.

Workover fluids are used during remedial work in the well such as removing tubing, replacing a pump, logging, reperforating, and cleaning out sand or other deposits.

Both workover and completion fluids are used in part to control the well pressure, to stop the well from blowing out while it is being completed or worked over, or to prevent the collapse of casing from over pressure.

Specially formulated fluids are used in connection with completion and workover operations to minimize the damage in the formation [2]. Formation damage from solids and filtrate invasion may be minimized by treating the well in a near-balanced condition, i.e., the wellbore pressure is close to the formation pressure. In high-pressure wells, it is often necessary to treat the well under over-balanced or under-balanced conditions. If over-balanced, the treating fluids are designed to temporarily seal the perforations to prevent the entry of fluids and solids into the formation. If under-balanced, the treating fluids are designed to prevent entry of solids from the formation into the wellbore.

A high-temperature treating fluid composition for well completion and workover is a saturated brine with parenchymal cell cellulose or a temperature-stable viscosifying bacterial cellulose product as suspension additive. Salts for brines are summarized in Table 53.

Potassium chloride and sodium chloride are preferred for low to moderate density fluids. For higher density fluids, cesium formate may be used.

A sized salt is used as a bridging agent in saturated salt water systems. A sized salt can seal a permeable zone by plugging the pores. It is a preferred

Water-Based Chemicals and Technology for Drilling, Completion, and Workover Fluids
http://dx.doi.org/10.1016/B978-0-12-802505-5.00004-4

Copyright © 2015 Elsevier Inc.
All rights reserved.

Table 53 Salts for brines [2]

Soluble salt	Sized salt
Potassium chloride	Potassium chloride
Sodium chloride	Sodium chloride
Calcium chloride	Calcium chloride
Sodium sulfate	Sodium sulfate
Sodium carbonate	Sodium carbonate
Sodium bicarbonate	Sodium bicarbonate
Calcium bromide	Calcium bromide
Sodium bromide	Sodium bromide
Potassium bromide	Potassium bromide
Magnesium bromide	Magnesium bromide
Magnesium chloride	Magnesium chloride
Potassium carbonate	Potassium carbonate
Potassium bicarbonate	Potassium bicarbonate
Potassium formate	Cesium chloride
Cesium formate	Cesium formate
Cesium chloride	Potassium formate

bridging agent because it can be dissolved by a water treatment with low salinity in order to eventually clean up the zone.

As water-soluble filtration additive a copolymer from 2-acrylamido-2-methylpropanesulfonate, acrylamide, or 2-vinylpryrrolidone can be used [2].

The wellbore fluid is effective for temporary bridging and plugging of productive formations so that cleanup can be effected by circulating connate water, field brine, or unsaturated brine solutions in the wellbore. Well completion operations can be performed at high borehole temperatures and pressures up to about 230 °C.

1.1 Logging Fluids

After a drilling process, a string of casing is lowered into the wellbore after the drill pipe is removed.

A logging fluid, such as a water-based mud, is used in the wellbore to compensate the formation pressure. A fill material, i.e., cement is then pumped into the annulus between the casing and the well wall. This replaces the mud and forms a cement bond, which isolates the formation layers and protects the casing [3].

The evaluation of the cement bond is important to check whether the bond functions properly to prevent the liquids from migrating from one formation layer to another. Such an evaluation is typically done during cement bond logging, using sonic or ultrasonic transmitters and sensors.

In the course of acoustic logging, acoustic pulses are emitted from the transmitters in a sonde. These pulses pass through the fluids inside the casing, and are partially reflected from the fluid-steel interface. A part of the pulses propagates further and is partially reflected at the steel-cement interface and at the cement-formation interface. The reflected signals are recorded by acoustic sensors and are analyzed.

The sensitivity depends on the impedance mismatching between steel casing and fluids in comparison to casing and cement. The impedance Z of a material 1 can be expressed as:

$$Z_1 = \rho_{0,1} C_1 \tag{2}$$

Here, $\rho_{0,1}$ is the static density of the material 1, C_1 is the speed of sound in the material 1.

The transmitted and reflected amplitude of acoustic waves at the interface between material 1 and material 2 of different impedances is given by:

$$A_i - A_r = \frac{Z_1 \sec \Theta_1}{Z_2 \sec \Theta_2} A_t \tag{3}$$

Here, A_i is the amplitude of the incident wave, A_r is the amplitude of the reflected wave, A_t is the amplitude of the transmitted wave, Θ_1 is the angle of propagation in the material 1, and Θ_2 is the angle of propagation in the material 2. In Table 54, the sound velocity C, the density ρ, and the Z factors for a number of materials are given.

It can be seen from Table 54 that steel and cement have very different Z values. The ratio of Z_{steel}/Z_{cement} is roughly 4.8. This ratio is significantly different from Z_{steel}/Z_{water}, which is 31.3. This difference is the base of cement bond logging. Therefore, by measuring the power reflected from the back side of the casing is a sensitive way to tell if the interface is steel/cement or steel/fluid, in other words, whether the cement has completely replaced the fluid [3].

By adding silica to a water-based mud, the ratio $Z_{steel}/Z_{watermud}$ still is significantly lowered. Thus, such a formulation enhances the sensitivity of the logging method. In contrast, the addition of cork increases the ratio. In this way, the formulation can be tailored accordingly [3].

Table 54 Acoustic properties of materials [3]

Material	C (m s^{-1})	ρ (g cm^{-3})	Z
Aluminum	6420	2.7	17,270
Beryllium	12,890	1.9	23,847
Steel, 1% C	5940	7.9	46,926
Titanium	6070	4.5	27,315
Silica	5968	2.6	15,756
Glass, flint	3980	2.6	10,348
Lucite	2680	1.2	3189
Nylon	2620	1.2	3144
Poly(ethylene)	1920	1.1	2112
Pentadecene	1351	0.78	1054
Water	1497	1.0	1497
Seawater	1535	1.1	1689
Cement	3200	2.9	9860
Cork	400	0.24	108
Barite	4000	4.2	1922

2. BOREHOLE REPAIR

In the course of drilling, the borehole often penetrates shallow water-bearing sand formations [4]. When the surface casing is subsequently installed in the well and cemented in place, the water in these sands may flow from behind the surface casing and washing out the sand grains or the cement, thus producing voids.

The washed out cement and the voids reduce the structural integrity of the casing. Such a reduction of the structural integrity results in an inability to keep the water flow and to support the wellhead equipment.

Consolidating incompetent formations penetrated by a borehole is done with special drilling fluids [4]. A water-based drilling fluid composition is used. The solids of the fluid are from a microfine-ground blast furnace slag. To prevent a flow from the water-bearing formations and to reduce the collapse of the incompetent formations, the drilling fluid should have a sufficient density to give a hydrostatic pressure greater than that of the formation fluid. This pressure difference causes the drilling fluid to flow into the formation, thus carrying the microfine blast furnace slag into the formation pores and building the filter cake.

Naturally, the drilling fluid density is increased by the addition of the blast furnace slag. If necessary, the density can be further increased by adding a soluble salt such as NaCl, CaCl$_2$, NaBr, ZnBr$_2$, or else standard insoluble weighting materials such as ground barite.

Table 55 Formulation of an in situ repair fluid [4]

Components	Unit	A	B
Seawater	bbl	1.0	1.0
NaCl	lbs bbl^{-1}	61.3	61.3
XC-Polymer 1 (water-soluble polymer)	lbs bbl^{-1}	1.5	1.5
FLR XL2 (water-soluble polymer)	lbs bbl^{-1}	1.5	1.5
CA-6003 (dispersant)	lbs bbl^{-1}	0.8	0.8
MC-1004 (blast furnace slag)	lbs bbl^{-1}	41.6	41.6
NaOH (activator)	lbs bbl^{-1}	4.0	–
Additional NaOH (after 16 h)	ppb	6.0	10.0

The addition of ground barite, unfortunately, will accelerate the filter cake buildup and so the penetration of the blast furnace slag particles is reduced. Therefore, a weight adjustment by using soluble salts is preferred. In this way, a fluid is obtained that will penetrate the incompetent formation and harden in place to consolidate the formation. A typical formulation is shown in Table 55.

As shown in Table 55, the drilling fluid may contain no activator since the fine particles of blast furnace slag which migrate into the formation will eventually harden due to the action of the water and salts present in the formation. A small amount of activator speeds up the hardening.

Subsequent cementing with a cementitious slurry results in a compatible cement, which will adhere to the filter cake formed from the drilling process.

3. CEMENTING

Compositions for cementing are usually water based. There are two basic kinds of activities in cementing, namely, primary and secondary cementing. Primary cementing fixes the steel casing to the surrounding formation. Secondary cementing is for filling formations, sealing, water shutoff, etc.

3.1 Primary Cementing

The main purposes of primary cementing are:
- Supporting vertical and radial loads to the casing,
- Isolating porous formations,
- Sealing subsurface fluid streams, and
- Protecting the casing from corrosion.

3.2 Secondary Cementing

Secondary cementing refers to cementing operations that are intended to use cement in maintaining or improving the operation of the well. There are two general cementing operations that are belonging to secondary cementing, i.e., squeeze cementing and plug cementing.

3.3 Squeeze Cementing

Squeeze cementing is used for the following purposes:
- Repairing a faulty primary cementing operation,
- Stopping intolerable loss of circulation fluid during drilling operations,
- Sealing abandoned or depleted formations,
- Repairing leaks of the casing, and
- Isolating a production zone by sealing adjacent unproductive zones.

The slurry should be designed to allow the fluid loss of the formation to be squeezed into the respective formation. Low-permeability formations can have a formulation of the slurry with an American Petroleum Institute (API) fluid loss [5] of 200-400 ml h^{-1}, whereas high-permeability formations require a slurry with 100-200 ml h^{-1} water loss. A high-pressure squeeze operation with a short duration requires an accelerator.

Thick slurries will not fill a narrow channel well. Therefore, squeeze cement slurries should be rather thin. Dispersants should be added for this reason. High compressive strength is not necessary for these types of slurries.

3.4 Plug Cementing

Plug cementing is used for plugging abandoned wells for environmental reasons. A kick-off plug is used to plug off a section of the borehole. The plug uses a hard surface to assist the kick-off procedure. Plug cementing is also used in drilling operations if extensive circulation loss is observed. The plug is set in the region of the thief zone and pierced again with the bit.

Often in open-hole completion operations and in production, it is necessary to shut off water flows. Additional cementation methods are used to provide an anchor for testing tools or for other maintenance operations.

3.5 Accelerators

Accelerators are important cementing additives for deepwater wells, since low temperatures can lengthen the setting time [6]. Traditionally used accelerators are inorganic salts such as calcium chloride. These accelerators may have a potentially negative side effect as they increase the permeability.

Nanosilica compounds have been shown to be good candidates as replacement materials for the traditional salt accelerators. The hydration of an oil-well cement may be accelerated by the addition of nanosilica; however, details have not been investigated.

It has been shown that the particle shape of the cement has an influence on the cement hydration kinetics. Smaller particle sizes and higher aspect ratios enhance the acceleration effect of nanosilicas [6].

3.6 Set Retarders

Set retarders are used to prevent a premature hardening of the cement slurry before it reaches the area to be cemented. Set retarders prolong the setting time of the cement to allow time for the cement to be pumped into place [7]. Common retarders used include lignins and sugars, as well as some metal oxides and acids.

Sometimes, the upper temperature limit of these retarders sometimes is too low for the cementing of high-temperature wells. So, the addition of a retarder enhancer is often required [8]. Sodium borate salts, e.g., borax and boric acid are known to be effective retarder enhancers. However, these chemicals are not always compatible with some other high-temperature additives and, therefore, may impair the fluid loss control and rheology of cement slurries. Poly(silicate)s have been proposed as alternatives [8].

To address environmental concerns, carboxymethylhydroxyethylcellulose, which is a biodegradable material, has been used in the past as a retardant in cement compositions [9]. Drawbacks encountered with carboxymethylated cellulose include very high surface slurry viscosities, particularly when higher quantities of the retarder are used to provide a retardation for cementing high-temperature zones. Other drawbacks to using carboxymethylated cellulose include a decrease in slurry viscosity due to thermal thinning as the slurry temperature increases to wellbore temperature during pumping and placement behind the casing, which may result in potential particle settling.

3.6.1 Seeds

Certain seeds can act as set retarders [7]. Suitable seeds are mustard seeds, navy beans, pinto beans, blackeye peas, popcorn, and dill seeds. Yellow or black mustard seeds are added in amounts of 0.1-2.0%.

3.6.2 Calcium Aluminate Cement Composition

It has been discovered that an organic acid and a polymeric mixture can function effectively as a set retarder for calcium aluminate cement compositions [10].

The organic acid is selected preferably from citric acid. The polymers are carboxymethyl cellulose and lignosulfonate. Other additives are a filler, a fluid loss additive, a friction reducer, a light-weight additive, a defoaming agent, a high-density additive, a mechanical property enhancing additive, a lost-circulation material, a filtration-control additive, and a thixotropic additive [10].

3.6.3 Polymer from Epoxysuccinic Acid

It has been discovered that a water-soluble biodegradable homopolymer of epoxysuccinic acid can be used as a set retarder [11]. Another advantage is that this polymer is a green scale inhibitor and it is biocompatible. For a fixed temperature, the thickening time increases with an increase in concentration of poly(epoxysuccinic acid).

NaCl acts as an accelerator for a cement composition when the concentration of the salt is less than 15% by weight of the water. But NaCl shows a retarding effect when the concentration of NaCl is in the range of 20-35% by weight of the water [11].

3.6.4 Carboxymethylated Inulin

A cement composition that contains carboxylated inulin, for example, carboxymethylated inulin, overcomes problems such as high temperatures that may result in particle settling [9]. For instance, carboxymethylated inulin provides the cement composition with a very low free water separation. In addition, the carboxymethylated inulin allows the cement composition to maintain its viscosity under increasing wellbore temperatures.

The method of carboxylating inulin consists of the carboxymethylation of inulin with chloroacetate to obtain the carboxymethylated inulin. The synthesis has been detailed [12].

Alternately, the carboxyl groups may be introduced into inulin by oxidation with suitable oxidants to form carboxylated inulin. Such oxidized inulins have been described in the literature [13]. The oxidation is achieved with periodic acid or periodate salts, lead(IV) salts, or permanganate.

3.7 Sealant Cement Composition

Squeezing or remedial cementing is a common operation in the petroleum industry [14]. Squeeze cementing is most often performed to repair leaks in well tubulars and to restore the pressure integrity to the wellbore or restore a cement sheath behind the casing. Squeeze cementing coupled with coiled tubing has been a standard remediation technique for shutting of unwanted gas or water production.

Cement is able to fill perforation tunnels, channels behind pipe, and/or washout zones behind pipe, and consequently cement is able to provide a near wellbore block to production.

Polymer gels are also used for shutting off of unwanted gas or water production. The main difference from squeeze cementing is that the polymer gels provide in depth blockage of the formation by penetrating the porous media and crosslinking in situ therein.

A limitation of the gels is that they may not have the mechanical properties to provide sufficient resistance to flow in the absence of a porous medium. A logical solution to these limitations is to combine polymer gels with cement squeezes to effectively block the production through perforations, voids, or cavities.

Such a combination is typically conducted sequentially. First, the polymer gel is placed in the formation and then the treatment is completed with a cement tail-in to squeeze the perforations behind pipe. A disadvantage of this sequential treatment may be that the depth of polymer invasion in the porous media extends beyond the depths that can be penetrated by perforating guns and consequently the shutoff may be permanent [14].

Another approach to combine the squeeze cementing and the polymer gel technology is to use the polymer gel as the mix water for the cement slurry [14].

The polymer is formed in situ from monomers such as acrylics. Hydroxyethylacrylate is the most preferred monomer [14].

The water-soluble polymerizable monomers are often used in combination with crosslinkable multifunctional vinyl monomers such as glycerol dimethacrylate and diacrylates. Suitable polymerization initiators can be sodium persulfate, or potassium persulfate and related compounds.

3.8 Gas Channeling Prevention

Methylhydroxyethyl cellulose may be used as an additive in cement slurries in the treatment of wells to prevent gas channeling [15]. Moreover, methylhydroxyethyl cellulose controls fluid loss, minimizes free fluid, and stabilizes foam during the cementing of the well.

3.9 Nano Clay

It has been recognized that the addition of a nano clay to cement compositions may have an impact on the physical characteristics of the cements such as compressive strength and tensile strength [16].

In addition, a nano clay may reduce the permeability of the final set cement. A significant decrease in permeability was observed for cement compositions that contain a nano bentonite as compared with regular bentonite. The permeability reduction ranged from 29% to 80%.

Examples of suitable nano clays include nano bentonite or nano montmorillonite. Nano montmorillonite is a member of the smectite clay family, and belongs to the general mineral group of clays with a sheet-like structure. It has a three-layered structure of aluminum sandwiched between two layers of silicon, similar to mica-type layered silicates.

Montmorillonite is an active and major ingredient in a volcanic ash called bentonite, which has an ability to swell to many times its original weight and volume when it absorbs water.

The nano clay may be encapsulated to facilitate the transportation and the incorporation into a well-treatment fluid [16].

4. FILTER CAKE REMOVAL

Conventional filter cake removal treatments use an array of chemicals specifically targeting two of the three components of a water-based filter cake [17]:

1. Chelants or acids to dissolve the calcite component, and
2. Enzymes or oxidizers to degrade the polymer component.

Well-treatment fluids for filter cake removal in gravel packing use a gravel carrying fluid containing enzyme for polymer removal in filter cake remediation, chelating agent to dissolve carbonate, and a viscoelastic surfactant [18]. However, these treatments fail to dissolve the third component of a water-based filter cake, i.e., the drilling solids and clays [17].

A three-step process to remove the filter cake by sequentially treating to remove each of the filter cake components has been described [19].

The removal of filter cake in a single step without the aid of a catalyst or activator as a separate completion step has been described [17]. This method includes the steps of: pumping a treatment fluid into the borehole in contact with the filter cake to be removed to establish a differential pressure between the treatment fluid and the formation adjacent the filter cake for a period of time. The treatment fluid contains an aqueous solution of a fluoride and another acid to provide an initial pH in the solution at 25 °C of 1.8-5. The fluoride source can be selected from the group consisting of ammonium fluoride, ammonium bifluoride, or polymeric fluorides. Ammonium bifluoride is the preferred compound.

If desired, the treatment fluid can also include an enzyme or oxidizer. Enzymes can include, for example, oxidoreductases, hydrolases, or lyases.

Mild oxidizers include ammonium persulfate, peroxides, or sodium bromate. The oxidizer should be present at a sufficiently low concentration to avoid a premature breakthrough from pinhole formation or non uniform removal of the filter cake.

5. NATURALLY FRACTURED CARBONATE RESERVOIR TREATMENT

In the Mexican Marine Region, gas breakthroughs are common in naturally fractured carbonate oil reservoirs. Increasing the gas production chokes also the production of crude oil [20].

In the Akal field, which is a large fractured carbonate reservoir, it has been shown that the gas–oil contact moves as fast as the natural gas. Thus, the injection gas moves through the natural fracture network and invades the oil zone. The result is declining production, depletion of the reservoir pressure, and oil remaining in the matrix.

A delayed crosslinked stable foamed fluid has been shown to be successful in selectively shutting off the gas in naturally fractured reservoirs. The fluid has a high foam quality and a low density.

The capillary pressure of the foam aids in displacing the gas and upward movement of the foam in the natural fissures into the gas cap. When the foam is cured, it forms an impermeable seal with a high extrusion resistance. After treatment, the oil production can be restored to the level prior to gas breakthrough [20].

6. WATER SHUTOFF

Unwanted water production is a serious issue in oil and gas producing wells. It causes corrosion, scale, and loss of productivity. One method of treating this problem is to chemically reduce unwanted water.

The usage of polymer systems for water shutoff and profile modification has been reviewed [21].

Field-application data for various polymer systems have been summarized over the range of 40-150 °C. These applications cover a wide range of permeabilities from 20 to 2720 mD in sandstone and carbonate reservoirs.

The first type is polymer gels for total water shutoff in the near-wellbore region, in which a polymer is crosslinked with either an organic or an inorganic crosslinking agent.

The second type is concerned with deep treatment of water injection wells diverting fluids away from high-permeability zones, i.e., thief zones. A thief zone collects most of the injected water, which results in a large amount of unrecovered oil.

Total-blocking gels are poly(urethane) resins, chromium crosslinking terpolymers [22], crosslinked foamed partially hydrolyzed poly(acrylamide) [23], and nanoparticle polyelectrolyte complexes [24]. Poly(ethylenimine) can crosslink various acrylamide-based polymers. Nanoparticles with effective diameters of about 100-200 nm are formed by a mixture of polyethylenimine and dextran sulfate in nonstoichiometric amounts [25].

A system that is applicable for a wide temperature range from 50 to 160 °C has been developed [26].

For the deep modification of water injection profiles in water injection wells, two systems are suitable: Microspheres prepared from acrylamide monomers crosslinked with N,N'-methylenebisacrylamide, as well as and microspheres using 2-acrylamido-2-methylpropane sulfonic acid as crosslinking agent [21].

Methods of water blocking or water shutoff in an oil or gas well have been described [27]. A crosslinked synthetic polymer gel formulation is used that is delivered into the well, whereby the formulation provides a water shutoff in the well.

A cationic poly(acrylamide) is used together with carboxymethyl cellulose. A gel can be produced by changing the pH. A synthetic polymer and dual crosslinking system, such as dialdehyde, e.g., glutaraldehyde, together with an organometallic reagent, such as zirconate and titanate complexes, is more stable to high-temperature conditions in comparison to only one crosslinking agent alone [27].

Other water-soluble polymers for oil field applications, among others for water shutoff are composed from both nonionic monomers and anionic monomers [28].

Table 56 Monomers for water-soluble polymers [28]

Nonionic monomers	Ionic monomers
Acrylamide	Dimethylaminoethyl acrylate
Methacrylamide	Dimethylaminoethyl methacrylate
N-Isopropyacrylamide	Dimethylaminopropyl acrylate
N,N-Dimethylacrylamide	Dimethylaminopropyl methacrylate
Diacetone acrylamide	Acrylic acid
N-Vinylformamide	Methacrylic acid
N-Vinyl acetamide	Itaconic acid
N-Vinylpyridine	Crotonic acid
N-Vinyl caprolactam	Maleic acid
N-Vinylpyrrolidone	Fumaric acid

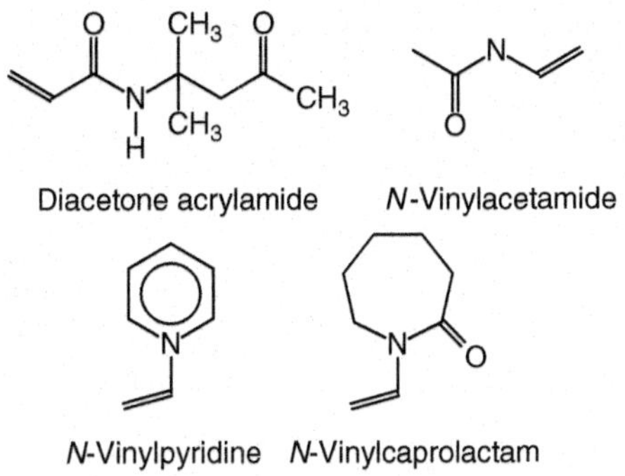

Diacetone acrylamide   N-Vinylacetamide

N-Vinylpyridine   N-Vinylcaprolactam

Figure 49 Monomers for water-soluble polymers.

The nonionic monomer are water-soluble vinyl monomers, and the ionic monomers are vinyl acids or tertiary amines. These monomers are listed in Table 56 and some compounds are shown in Figure 49.

7. SWEEP FLUIDS

Cuttings transport from the bit to the surface in horizontal or highly inclined wells is a critical component of the overall drilling performance [29]. Namely, the associated increase in bottomhole pressure and equivalent circulating density due to increased cuttings concentration and formation of cuttings bed in the wellbore can exceed the limits of the marginal operating pressure window that often exists in deepwater drilling.

Fiber-containing sweeps, or *fiber sweeps,* have been shown to be effective in cleaning highly inclined wells by eliminating cuttings beds, thereby reducing the friction pressure loss. Incorporating a fiber in a drilling sweep enhances its hole-cleaning performance and minimizes friction loss contributions from cuttings beds without significantly impacting flow properties of the fluid.

In an inclined configuration, the addition of fibers to hole-cleaning sweeps substantially improves cuttings removal, but only when the pipe is rotated. When the annulus is horizontal or the pipe is not rotated, addition of fiber has only a small effect on cuttings bed removal [29].

8. WATER-BASED ENHANCED OIL RECOVERY

In mineral oil production, a distinction is drawn between primary, secondary, and tertiary production [30].

In primary production, after commencement of drilling, the mineral oil flows to the surface owing to the autogenous pressure of the deposit. The autogenous pressure can be caused, by gases present in the deposit such as methane, ethane, or propane. In this stage, only 5-10% of the oil in place can be produced.

Typically, after primary production, secondary production is used. Here, additional boreholes, so-called injection boreholes are drilled. In these holes, water is injected in order to maintain or to increase the pressure. In the second stage, some 30-35% of the oil in place can be produced.

Afterwards a tertiary mineral oil production, also addressed as enhanced oil recovery (EOR) is used. An overview of tertiary oil production using chemicals can be found in the literature [31, 32].

Petroleum recovery is typically accomplished by drilling into a petroleum-containing formation and utilizing one of the well-known methods for the production of petroleum [33]. However, these recovery techniques may recover only a minor portion of the petroleum present in the formation particularly when applied to formations containing viscous petroleum.

In such cases, secondary recovery methods, such as waterflooding, steam injection, gas flooding, and combinations thereof, may be used to enhance petroleum recovery. Underground oil-containing formations also contain clay or clay-like bodies and treatment with water or steam generally results in swelling of the clay by absorption of the water, with the result that the water permeability of a formation is decreased. The decrease in the permeability of the formation causes a reduction in the amount of oil which may be recovered by secondary recovery operations.

The production of petroleum may be improved or reinstated from formations which have been partially depleted by primary recovery techniques.

Such methods include the injection of an aqueous composition into a first well to force residual petroleum in an underground formation through the formation and out of one or more recovery wells. A water dispersible polymeric material such as a hydrated polysaccharide may be included in the aqueous composition to increase the viscosity, when the residual petroleum is viscous.

In general, the injected aqueous compositions have a lower viscosity at reservoir conditions than the viscosity of the formation crude which it is intended to displace, making it less effective. Various additives, such as polymers, have been proposed to increase the viscosity of the injected fluid in order to improve the efficiency [33].

The composition of a brine to be injected has a profound effect on the efficiency of water-based EOR methods [34].

It has been observed that not only is the concentration of the active ions Ca^{2+}, Mg^{2+}, and SO_4^{2-} important for wettability alteration in carbonates, but also the amount of non active salt of NaCl, has an impact on the oil recovery process. Removing NaCl from synthetic seawater improves the oil recovery by about 10% of original oil in place (OOIP) compared to ordinary seawater. The results have been discussed in terms of electrical double-layer effects. The seawater was depleted in NaCl by adjusting the concentration of the active ions, Ca^{2+} and SO_4^{2-}. Oil displacement studies in outcrop chalk samples by spontaneous imbibition were performed at temperatures of 70-120 °C using different oils and imbibing fluids. When the concentration of SO_4^{2-} in the seawater depleted with regard to NaCl was increased 4 times, the ultimate oil recovery increased by about 5-18% of OOIP in comparison to the seawater depleted in NaCl. The amount of Ca^{2+} in the seawater depleted in NaCl had no significant effect on the oil recovery at 100 °C, but significant improvements were observed at 120 °C. A chromatographic wettability analysis showed that the water-wet area of the rock surface increased, as the oil recovery increased. Thus, the importance of the ionic composition and the ion concentration of the injecting brine in the water–based EOR methods has been demonstrated [34].

pH responsive amphiphilic systems can be used as displacement fluids in EOR [35]. The complexation and the supramolecular assembly of an amino amide and maleic acid is able to control the viscosity of aqueous displacement fluids.

It has been shown that the addition of only 2% of the composition into water increases its viscosity by a factor of 4.5×10^5. This superior behavior has been attributed to the formation of layered cylindrical supramolecular assemblies. In addition, the viscosity can be increased 12 times by increasing the pH. This is a reversible process [35].

To improve the sweep efficiency for carbon dioxide EOR up to 120 °C in the presence of high-salinity brine carbon dioxide/water foams have been formed with surfactants composed of ethoxylated amine headgroups with cocoalkyl tails [36]. These surfactants are switchable from the nonionic state in dry carbon dioxide to a cationic state in the presence of an acidic aqueous phase with a pH less than 6.

The foams were produced by injecting the surfactant into either the carbon dioxide phase or the brine phase. This behavior indicates a good contact between the two phases. The switchable ethoxylated alkyl amine surfactants exhibit both high cloudpoints in brine and high interfacial activities in the ionic state in water for foam generation [36].

8.1 Waterflooding

The most common form of secondary recovery is the process known as waterflooding [37]. In a waterflood project, water is injected into the oil-producing formation through injection wells, repressurizing the formation and sweeping oil which would not have otherwise been produced into production wells. Such a procedure will usually allow the economic production of an additional 10-30% of the oil originally in place.

8.1.1 Aqueous Surfactant System

It has been proposed to add surfactants to the flood water in order to lower the oil-water interfacial tension or to alter the wettability characteristics of the reservoir rock [38]. Processes which involve the injection of aqueous surfactant solutions are commonly referred to as surfactant waterflooding or as low tension waterflooding.

Many waterflooding processes have used anionic surfactants. One problem encountered in waterflooding with certain of the anionic surfactants such as the petroleum sulfonates is the lack of stability of these surfactants in so-called *hard water* environments.

These surfactants tend to precipitate from solution in the presence of relatively low concentrations of divalent metal ions such as calcium and magnesium ions. For example, divalent metal ion concentrations of about 50-100 ppm and above usually tend to cause precipitation of the petroleum sulfonates.

On the other hand, nonionic surfactants, such as polyethoxylated alkyl phenols, polyethoxylated aliphatic alcohols, carboxylic esters, carboxylic amides, and poly(oxyethylene) fatty acid amides, have a higher tolerance to polyvalent ions such as calcium or magnesium than anionic surfactants.

However, nonionic surfactants are not as effective than anionic surfactants and nonionic surfactants generally are more costly. In addition, nonionic surfactants may show a reverse solubility relationship with temperature and thus become insoluble at temperatures of above their cloud points. On the other hand, nonionic surfactants that remain soluble at elevated temperatures are generally not effective in reducing the interfacial tension. Moreover, nonionic surfactants usually hydrolyze at temperatures above 75 °C [38].

The use of amphoteric surfactants which function as cationics in acid media and become anionic when incorporated in alkaline systems has been suggested. An EOR method that uses betaine amphoteric surfactants in combination with high molecular weight homopolysaccharide gum thickeners in a waterflood process has been described [39]. The waterflood can be followed by a thickened buffer slug and then an aqueous flooding medium to displace the oil toward a production well.

8.1.2 Aqueous Surfactant System and Kraft Lignin

A method for the enhanced recovery of oil is to inject an aqueous surfactant system into an oil-containing formation [37]. This system contains an unmodified Kraft lignin, an oil-soluble amine, and a water-soluble sulfonate, as well as water. The surfactant system is driven through the formation and the oil mobilized by the surfactant system is then produced.

The amine is a primary fatty amine, e.g., tallow amine, and the anionic surfactant is an alkyl-aryl sulfonate [37]. Fatty amines are highly insoluble in water and tend to precipitate when water is added. However, under proper conditions, the amines can be dissolved in water which contains a surfactant such as a petroleum sulfonate or an alkyl-aryl sulfonate together with the lignin.

A suitable mixing procedure is to combine the amine and the lignin and then to add brine which has been before preheated to a temperature above the melting point of the amine. In the case of tallow amine, a temperature of about 60-70 °C is adequate. This mixture is then stirred at about 65 °C.

for 1 h. Then a water-soluble surfactant is then added directly into the warm solution. After additional stirring for 1-5 h at 65 °C the solution is allowed to cool.

Also a procedure to get water-soluble surfactants from lignin by the modification of lignin has been described [40] This consists of alkylating the lignin at its phenolic oxygen sites, sulfonating the alkylated lignin, and oxidizing the so modified lignin to break the molecule into smaller compounds that are water soluble and can act as surfactants. These oxidized lignin surfactants in surfactant floods can be used for EOR.

8.2 Brine Tolerant Foam Composition

An EOR technique has been reported that uses a periodic injection of gas and a foam-forming composition into the reservoir.

A typical procedure involves the injection of a slug of CO_2 followed by the injection of a higher viscosity fluid such as water to push the CO_2.

The foam-forming composition contains water, a sulfonate surfactant as foam-forming compound and a solubilizing component [41].

Sodium xylene sulfonate is such a sulfonate surfactant, among others. Solubilizers are compounds which are not suitable foaming agents but they can improve brine tolerance of less brine tolerant materials [42]. Materials which have been employed as solubilizers include nonionic surfactants such as ethoxylated nonylphenols and secondary alcohols, ethylene/propylene oxide copolymers, fatty acid ethanolamides, glycols, and polyglycosides as well as anionic, cationic, and amphoteric surfactants. Solubilizing components have been examined in detail for a brine containing both sodium and calcium ions [43].

8.3 Polymer Flooding

One approach to tertiary oil recovery is polymer flooding [44]. There, certain organic polymers are induced into the formation to thicken the fluid and thereby to improve the mobility of the entrapped oil as the fluid is driven from the injection site to the production well.

8.4 Acrylic Polymers

For polymer flooding, a multitude of different thickening polymers have been proposed, especially high molecular weight poly(acrylamide), copolymers of acrylamide and further comonomers, for example, vinylsulfonic acid or acrylic acid [30, 45]. Poly(acrylamide) may especially be partly hydrolyzed polyacrylamide, in which some of the acrylamide units have been hydrolyzed to acrylic acid.

Hydrophobically associating copolymers can also be used for polymer flooding. These are water-soluble polymers which have lateral or terminal hydrophobic groups, for example, relatively long alkyl chains.

In aqueous solution, such hydrophobic groups can associate with themselves or with other substances having hydrophobic groups. This forms an associative network by which the medium is thickened.

Details of the use of hydrophobically associating copolymers for tertiary mineral oil production have been described [46].

A mixture of hydrophobically associating polymers and surfactants for tertiary mineral oil extraction have been described [47]. Hydrophilic comonomers are acrylamides. The hydrophobic monomers are long chain alkyl esters of unsaturated carboxylic acids, vinyl alkylates, and alkylstyrenes.

Specifically, copolymers from acrylamide and 2-acrylamido-2-methylpropane sulfonic acid have been described [30].

8.5 Naturally Occurring Polymers

In addition, it is also possible to use naturally occurring polymers. Natural derived polymers include xanthan gums, hydrophilic polysaccharides which are produced by a fermentation using bacteria of the genus *Xanthomonas*. The biopolymer may be used in the form of the fermentation broth itself or in isolated and reconstituted form [44, 48].

Xanthan gums are particularly suitable for polymer flooding, since they are good displacing agents, result in useful viscosities even at low concentrations. Further, they are not lost by extensive adsorption of the porous rock formations and are relatively insensitive to salts [44].

Xanthan gums are susceptible to chemical degradation or depolymerization. The degradation, which tends to increase with increasing temperature, reduces the viscosity of a solution containing the polymers. Two paths by which such a degradation can occur are either hydrolysis or free-radical reactions.

The hydrolysis reaction is the reaction of water with the ether-type linkages in the structure of the polysaccharide. Free-radical reactions are usually initiated when the polymer solution is mixed with air or oxygen. Such a mixture tends to form hydroperoxides and the decomposition of the hydroperoxides produces reaction-initiating free-radicals that propagate polymer-degrading radical reactions [49].

The stability of the polysaccharide solution is increased by the addition of an alkali metal borohydride, which prevents an oxidation reaction [44]. Further, a water-soluble sulfur containing antioxidant is capable of stabilizing a Xanthan gum polymer solution [49].

Nevertheless, it is desirable to have a low dissolved oxygen content when the stabilizer is used, and this is most readily accomplished by use of separated recycled brine produced from the reservoir under anaerobic conditions. Natural gas or other cheap inert gas blankets may be used in surface handling to maintain the anaerobic conditions [44].

8.5.1 Continuous Dissolution of Polymer Viscosifier

An EOR method has been described that consists of continuously dissolving in the injection water a stable invert emulsion of an acrylamide copolymer, an inverting agent, and a water-soluble polymer [50].

Acrylamide polymers are produced industrially as powders, water-in-oil emulsions, polymerized beads in suspension, water-in-water emulsions, aqueous solutions, and water-alcohol suspensions. These polymers may have similar physicochemical properties and are usually selected on two criteria: Cost and ease of handling. Mostly, despite higher costs, emulsions are selected because of ease of handling.

Such emulsions have been already described in 1957 [51]. These early emulsions contained microparticles of polymers emulsified in an oil by means of one or more surface active agents having a low hydrophilic-hydrophobic balance of 2-6. For these emulsions having such composition, it was very difficult to dilute the polymer microgels directly in a solution.

In 1970, it was discovered that it was possible to invert these emulsions, by adding a hydrophilic surface active agent [52]. This rather hydrophilic surface active agent enabled the water to penetrate into the polymer spherical microparticle and to dissolve it in a few minutes.

Initially, this surface active agent was delivered and added separately to the dilution water [50]. But very soon, the production companies developed formulations in which it was possible to incorporate inverting surface active agents in the emulsion itself, without latter coagulating or being destabilized.

This operation was a real technical challenge because required quantity of hydrophilic surface active agent added was limited by coagulation or destabilization either during production or in storage. The industrial implementation of this method was the subject of considerable successive researches and numerous patents.

However, even today, limitations to properly invert an emulsion exist, with two main factors: The quantity of hydrophilic surface active agent the system can sustain to avoid coagulation of the emulsion during production or in storage, and the concentration of polymer reached when dissolving the emulsion in water.

It is commonly found that the maximum quantity of hydrophilic surface active agent, all systems combined, compatible with the stability of the emulsion, is about 5% of the emulsion [50]. For these reasons, the continuous dissolution of the polymer viscosifier has been developed [50].

To obtain a maximum viscosity a poly(acrylamide) copolymer type is used from 70% of acrylamide and 30% of acrylic acid. In a first step, the emulsion is prediluted in a first static mixer mounted on a bypass of the main injection water circuit, to a concentration of the polymer of $20\,g\,l^{-1}$, with a difference of pressure between the mixer outlet and the inlet of 10 bar. Then, in a second step, the dispersion mixture from the first mixer is diluted in a second static mixer that is mounted on the main injection water circuit to 1000-2000 ppm, with a difference of pressure of 3 bar.

8.5.2 Crosslinked Vinylamine Polymer

Polymers are used in acidizing or fracture acidizing in which acidic compositions are used to stimulate production of hydrocarbon from underground formations by increasing the formation porosity [53]. A water-soluble or water-dispersible polymer is incorporated to increase the viscosity of the fluid so that wider fractures can be developed and live acid can be forced farther into the formations. This increases the proppant carrying capacity of the acid solutions and permits better fluid loss control.

Unfortunately, certain polymeric viscosifiers are degraded by the hostile reservoir environment including high temperatures, acidity, and extreme shear conditions, as well as by the electrolytes which are encountered in the oil recovery process. For example, hydrolyzed polyacrylamides fail in sea water solution at elevated temperatures due to precipitation of the polymer in the presence of calcium ions in the sea water. Xanthan polymers are insensitive to calcium ions but these polymers degrade at high temperatures and lose their viscosifying efficiency.

It has been found that vinylamine polymers over a broad range of molecular weights can be crosslinked and used effectively due to their stability under harsh environmental conditions in EOR [53]. Such polymers

Table 57 Crosslinking agents for poly(vinylamine) [53]

Compound	pH	T (°C)	Gel time (min)
Glyoxal	6	25	0
Glyoxal	11	25	0
Glutaraldehyde	3	25	0
Glutaraldehyde	11	25	0
Diisocyanatohexane	5	25	30
Epichlorohydrin	11	25	30
Epichlorohydrin	9	90	10

exhibit a good stability at high temperatures under acid conditions and high electrolyte concentrations and are particularly suitable for use in acid or matrix fracturing of oil or gas bearing formations.

The preferred method of synthesizing poly(vinylamine)s is by the hydrolysis of a poly(N-vinylformamide). Also copolymers from vinylacetate and N-vinylformamide can be used. On hydrolysis a copolymer containing poly(vinyl alcohol) linkages and poly(vinylamine) linkages is formed.

The vinylamine polymer can be crosslinked with either a multi-functional organic compound or an inorganic compound containing multivalent anions such as titanates, zirconates, phosphates, silicates, or an inorganic divalent cations that are capable of complexing with the vinylamine polymer. Organic crosslinking agents which are particularly advantageous are diepoxides and diisocyanates [53]. The crosslinked vinylamine polymer greatly increases the viscosity and may form a gel. Crosslinking agents are detailed in Table 57.

9. PIPELINES

Many different methods for cleaning pipeline networks have developed [54]. Cleaning methods in order to decrease and remove the iron and iron sulfide deposits include mechanical pigging, batch chemical cleaning, and continuous chemical cleaning. The chemical compounds used in batch and in continuous cleaning methods are surfactants, solvents, acids, bases, oxidizing agents, and chelating agents.

The use of a strong acid is the simplest way to dissolve such a deposit. But using a strong acid generates large volumes of highly toxic hydrogen sulfide gas, which is an undesirable byproduct. Adding an oxidizing agent may avoid such toxicity hazards but oxidation products are then produced, including elemental sulfur which is corrosive to pipes.

Table 58 Synergistic chelating agents [54]

Compound	Acronym
Polyaspartates	
Hydroxyaminocarboxylic acid	HACA
Hydroxyethyliminodiacetic acid	HEIDA
Iminodisuccinic acid	IDS
Ethylene diaminetetracetic acid	EDTA
Diethylenetriaminepentaacetic acid	DTPA
Nitrilotriacetic acid	NTA
Tetrakis(hydroxymethyl)phosphonium sulfate	THPS

Another agent for treating such deposits is acrolein, but it also has health, safety, and environmental problems. Tris(hydroxymethyl) phosphine can solubilize iron and iron sulfide by forming a bright red water-soluble complex.

A blend of chelating agents has been described that show a synergistic action. These chelating agents are summarized in Table 58 and some are shown in Figure 50.

Both iminodisuccinic acid and tetrakis(hydroxymethyl)phosphonium sulfate are environmentally friendly chelating agents [54].

It has been observed that the overall performance of the iron complexing agents improve at alkaline pH. The pH of such a mixture can be adjusted into the alkaline range using ammonium hydroxide, ammonium chloride, ammonium citrate, ammonium lactate, ammonium acetate, potassium citrate, potassium hydroxide, potassium formate, sodium hydroxide, sodium acetate, or sodium formate.

Also, metal ions other than iron can form soluble complexes and be removed in the cleaning process. The presence of a surfactant helps in the dispersion and avoids deposits from reforming at downstream points within a pipeline network [54].

9.1 Gelled Pigs

Gelled fluid pigs will perform most of the functions of conventional pigs, but they have additional chemical capabilities. In addition, they can be injected into a pipeline through a valve. However, for displacement by a gas, gel pigs must be propelled by a mechanical pig.

Most pipeline gels are water based, but a variety of chemicals, solvents, and acids can be gelled. Gelled diesel, an organic gel, was first patented for pipeline use in 1973 [55]. The gels can be used for waste material removal,

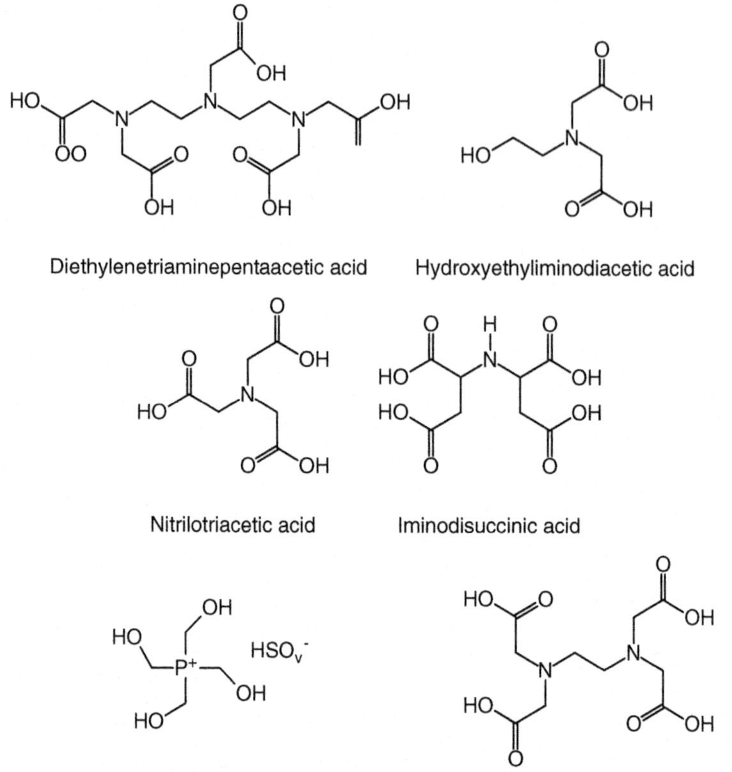

Diethylenetriaminepentaacetic acid   Hydroxyethyliminodiacetic acid

Nitrilotriacetic acid   Iminodisuccinic acid

Tetrakis(hydroxymethyl)phosphonium sulfate   Ethylene diaminetetracetic acid

Figure 50 Chelating agents.

separation of products, placement of biocides and inhibitors, and removal of trapped mechanical pigs [56, 57].

An ablating gelatin pig has been described for use in pipelines. Because of the properties of gelatin, the pig will ablate, thereby depositing a protective layer onto the wall of the pipe [58]. The pig can be molded outside the pipe or it can be formed in situ.

The pig is formed by mixing gelatin with a heated liquid and then allowing the mixture to cool to ambient temperature. Preferably, the liquid contains a corrosion inhibitor or a drag reducer. In some applications, a slug of the treating solution is also passed through the pipeline between two ablating gelatin pigs. For high-temperature applications, a hardener may be added to increase the melting temperature of the pig.

Tradenames appearing in the references are shown in Table 59.

Table 59 Tradenames in references

Tradename Description	Supplier
Baysilon® Silicone oil [30]	Bayer AG
Dowfax® (Series) Nononic surfactants [41]	Dow Chemical Comp.
MontBrite® 1240 Aqueous solution of sodium borohydride and caustic soda [33]	Montgomery Chemicals
Proban® Tetrakis(hydroxymethyl)phosphonium salts [54]	Rhodia
Pusher® (Series) Partially hydrolyzed polyacrylamides [47]	Dow Chemical Comp.
Surfynol® (Series) Ethoxylated acetylene diols [30]	Air Products and Chemicals, Inc.

REFERENCES

[1] Patel BB. Composition and process for stabilizing viscosity or controlling water loss of polymer-containing water based fluids. US Patent 5576271, assigned to Phillips Petroleum Company (Bartlesville, OK); 1996. URL: http://www.freepatentsonline.com/5576271.html.

[2] Elward-Berry J. Water based high temperature well servicing composition and method of using same. US Patent 5620947, assigned to Exxon Production Research Company (Houston, TX); 1997. URL: http://www.freepatentsonline.com/5620947.html.

[3] Freeman MA. Well logging fluid for ultrasonic cement bond logging. US Patent 8186432, assigned to M-1 LLC (Houston, TX); 2012. URL: http://www.freepatentsonline.com/8186432.html.

[4] Nahm JJW, Wyant RE. Well fluid for in-situ borehole repair. US Patent 5309997, assigned to Shell Oil Company (Houston, TX); 1994. URL: http://www.freepatentsonline.com/5309997.html.

[5] API Standard RP 13B-1. Standard procedure for field testing water- based drilling fluids. API Standard API RP 13B-1. Washington, DC: American Petroleum Institute; 1997.

[6] Pang X, Boul PJ, Cuello Jimenez W. Nanosilicas as accelerators in oilwell cementing at low temperatures. SPE Drill Complet 2014;29(01):98–105. doi:10.2118/168037-pa.

[7] Spangle L. Use of seeds as a cement set retarder. US Patent 8598091, assigned to Catalyst Partners, Inc. (Chico, TX); 2013. URL: http://www.freepatentsonline.com/8598091.html.

[8] Caritey JP, Michaux M, Pyatina T, Thery F. Versatile additives for well cementing applications. US Patent 7946343, assigned to Schlumberger Technology Corporation (Sugar Land, TX); 2011. URL: http://www.freepatentsonline.com/7946343.html.

[9] Santra AK, Reddy BR, Brenneis DC. Biodegradable retarder for cementing applications. US Patent 8435344, assigned to Halliburton Energy Services, Inc. (Duncan, OK); 2013. URL: http://www.freepatentsonline.com/8435344.html.

[10] Joseph T, Chakraborty PP, Melbouci M. Calcium aluminate cement composition containing a set retarder of an organic acid and a polymeric mixture. US Patent

8720563, assigned to Halliburton Energy Services, Inc. (Houston, TX); 2014. URL: http://www.freepatentsonline.com/8720563.html.

[11] Tarafdar A, Senapati D, Sarap GD, Patil RC. Epoxy acid based biodegradable set retarder for a cement composition. US Patent 8536101, assigned to Halliburton Energy Services, Inc. (Houston, TX); 2013. URL: http://www.freepatentsonline.com/8536101.html.

[12] Verraest DL, Batelaan JG, Peters JA, van Bekkum H. Carboxymethyl inulin. US Patent 5777090, assigned to Akzo Nobel NV (Arnhem, NL); 1998. URL: http://www.freepatentsonline.com/5777090.html.

[13] Veelaert S, De Wit D, Tournois H. Method for the oxidation of carbohydrates. US Patent 5747658, assigned to Instituut Voor Agrotechnologisch Onderzoek (ATO-DLO) (Wageningen, NL); 1998. URL: http://www.freepatentsonline.com/5747658.html.

[14] Dalrymple ED, Eoff LS, van Batenburg DW, van Eijden J. Sealant composition comprising a gel system and a reduced amount of cement for a permeable zone downhole. US Patent 8703659, assigned to Halliburton Energy Services, Inc. (Houston, TX); 2014. URL: http://www.freepatentsonline.com/8703659.html.

[15] Bray WS, Brandl A. Use of methylhydroxyethyl cellulose as cement additive. US Patent 8689870, assigned to Baker Hughes Incorporated (Houston, TX); 2014. URL: http://www.freepatentsonline.com/8689870.html.

[16] Roddy CW, Covington RL, Chatterji J, Brenneis DC. Cement compositions and methods utilizing nano-clay. US Patent 8603952, assigned to Halliburton Energy Services, Inc. (Houston, TX); 2013. URL: http://www.freepatentsonline.com/8603952.html.

[17] Pirolli L, Parlar M. Filtercake removal composition and system. US Patent 8114817, assigned to Schlumberger Technology Corporation (Sugar Land, TX); 2012. URL: http://www.freepatentsonline.com/8114817.html.

[18] Tibbles RJ, Parlar M, Chang FF, Fu D, Davison JM, Morris EWA, et al. Fluids and techniques for hydrocarbon well completion. US Patent 6638896, assigned to Schlumberger Technology Corporation (Sugar Land, TX); 2003. URL: http://www.freepatentsonline.com/6638896.html.

[19] Parlar M, Brady M, Morris L. Method for removing filter cake from injection wells. US Patent 6978838, assigned to Schlumberger Technology Corporation (Sugar Land, TX); 2005. URL: http://www.freepatentsonline.com/6978838.html.

[20] Lozada M, Torres M, Gonzalez M, Garcia B, Cortes M, Milne A, et al. Selectively shutting off gas in naturally fractured carbonate reservoirs. In: Naturally-fractured reservoirs. SPE International Symposium and Exhibition on Formation Damage Control. Lafayette, LA: Society of Petroleum Engineers; 2014, URL: https://www.onepetro.org/conference-paper/SPE-168195-MS.

[21] El-Karsani KSM, Al-Muntasheri GA, Hussein IA. Polymer systems for water shutoff and profile modification: a review over the last decade. SPE J 2014;19(01):135-49. doi:10.2118/163100-pa.

[22] Bjørsvik M, Høiland H, Skauge A. Formation of colloidal dispersion gels from aqueous polyacrylamide solutions. Colloids Surf A Physicochem Eng Asp 2008;317 (1-3):504-11. doi:10.1016/j.colsurfa.2007.11.025.

[23] Al-Assi A, Willhite G, Green D, McCool C. Formation and propagation of gel aggregates using partially hydrolyzed polyacrylamide and aluminum citrate. SPE J 2009;14(3). doi:10.2118/100049-pa.

[24] Cordova M, Cheng M, Trejo J, Johnson SJ, Willhite GP, Liang JT, et al. Delayed HPAM gelation via transient sequestration of chromium in polyelectrolyte complex nanoparticles. Macromolecules 2008;41(12):4398-404. doi:10.1021/ma800211d.

[25] Johnson S, Trejo J, Veisi M, Willhite GP, Liang JT, Berkland C. Effects of divalent cations, seawater, and formation brine on positively charged polyethylenimine/dextran sulfate/chromium(III) polyelectrolyte complexes and partially hydrolyzed polyacrylamide/chromium(III) gelation. J Appl Polym Sci 2010;115(2):1008-1014. doi:10.1002/app.31052.

[26] Al-Muntasheri GA, Sierra L, Garzon FO, Lynn JD, Izquierdo GA. Water shut-off with polymer gels in a high temperature horizontal gas well: a success story. In: SPE Improved Oil Recovery Symposium; 2010. doi:10.2118/129848-ms.

[27] Wuthrich P, Mahoney Rp, Soane Ds, Casado Portilla R. Crosslinked synthetic polymer gel systems for hydraulic fracturing. US Patent Application 20140158355, assigned to Soane Energy, LLC; 2014. URL: http://www.freepatentsonline.com/20140158355.html.

[28] Favero C, Gaillard N, Marroni D. Water-soluble polymers for oil recovery. US Patent Application 20130072405; 2013. URL: http://www.freepatentsonline.com/20130072405.html.

[29] George M, Elgaddafi R, Ahmed R, Growcock F. Performance of fiber-containing synthetic-based sweep fluids. J Petrol Sci Eng 2014;119:185-95. URL: http://www.sciencedirect.com/science/article/pii/S0920410514001260. doi:10.1016/j.petrol.2014.05.009.

[30] Reichenbach-Klinke R, Langlotz B. Aqueous formulations of hydrophobically associating copolymers and surfactants and use thereof for mineral oil production. US Patent 8752624, assigned to BASF SE (Ludwigshafen, DE); 2014. URL: http://www.freepatentsonline.com/8752624.html.

[31] Kessel DG. Chemical flooding—status report. J Petrol Sci Eng 1989;2(2-3):81-101. URL: http://www.sciencedirect.com/science/article/pii/0920410589900569. doi:10.1016/0920-4105(89)90056-9.

[32] Santanna V, Curbelo F, Dantas TC, Neto AD, Albuquerque H, Garnica A. Microemulsion flooding for enhanced oil recovery. J Petrol Sci Eng 2009;66(3-4):117-20. URL: http://www.sciencedirect.com/science/article/pii/S0920410509000588. doi:10.1016/j.petrol.2009.01.009.

[33] Lumsden CA, Diaz RO. Method and composition for oil enhanced recovery. US Patent 8662171, assigned to Montgomery Chemicals, LLC (Conshohocken, PA) Nalco Company (Naperville, IL); 2014. URL: http://www.freepatentsonline.com/8662171.html.

[34] Fathi SJ, Austad T, Strand S. water based enhanced oil recovery (EOR) by "smart water": optimal ionic composition for EOR in carbonates. Energy Fuels 2011;25(11):5173-9. doi:10.1021/ef201019k.

[35] Chen IC, Yegin C, Zhang M, Akbulut M. Use of pH-responsive amphiphilic systems as displacement fluids in enhanced oil recovery. SPE J 2014a. doi:10.2118/169904-pa.

[36] Chen Y, Elhag AS, Poon BM, Cui L, Ma K, Liao SY, et al. Switchable non-ionic to cationic ethoxylated amine surfactants for CO_2 enhanced oil recovery in high-temperature, high-salinity carbonate reservoirs. SPE J 2014b;19(02):249-59. doi:10.2118/154222-pa.

[37] Kieke DE. Use of unmodified Kraft lignin, an amine and a water-soluble sulfonate composition in enhanced oil recovery. US Patent 5911276, assigned to Texaco Inc. (White Plains, NY); 1999. URL: http://www.freepatentsonline.com/5911276.html.

[38] Kalpakci B, Arf TG. Surfactant-polymer composition and method of enhanced oil recovery. US Patent 5076363, assigned to The Standard Oil Company (Cleveland, OH); 1991. URL: http://www.freepatentsonline.com/5076363.html.

[39] Kalpakci B, Chan KS. Method of enhanced oil recovery employing thickened ampho-teric surfactant solutions. US Patent 4554974, assigned to The Standard Oil Company (Cleveland, OH); 1985. URL: http://www.freepatentsonline.com/4554974.html.

[40]   Morrow LR. Enhanced oil recovery using alkylated, sulfonated, oxidized lignin surfactants. US Patent 5094295, assigned to Texaco Inc. (White Plains, NY); 1992. URL: http://www.freepatentsonline.com/5094295.html.

[41]   Sevigny WJ, Kuehne DL, Cantor J. Enhanced oil recovery method employing a high temperature brine tolerant foam-forming composition. US Patent 5358045, assigned to Chevron Research and Technology Company, a Division of Chevron U.S.A. (San Francisco, CA); 1994. URL: http://www.freepatentsonline.com/5358045.html.

[42]   Allured M. McCutcheon's functional materials. Glen Rock, NJ: McCutcheon's Division, MC Pub. Co.; 2006.

[43]   Oswald T, Robson IA. Gas flooding processing for the recovery of oil from subterranean formations. US Patent 4860828, assigned to The Dow Chemical Company (Midland, MI); 1989. URL: http://www.freepatentsonline.com/4860828.html.

[44]   Philips JC, Tate BE. Stabilizing polysaccharide solutions for tertiary oil recovery at elevated temperature with borohydride. US Patent 4458753, assigned to Pfizer Inc. (New York, NY); 1984. URL: http://www.freepatentsonline.com/4458753.html.

[45]   Reichenbach-Klinke R, Langlotz B, Macefield IR, Spindler C. Process for tertiary mineral oil production. US Patent Application 20140131039, assigned to Basf se, Ludwigshafen (DE); 2014. URL: http://www.freepatentsonline.com/20140131039. html.

[46]   Taylor KC, Nasr-El-Din HA. Water-soluble hydrophobically associating polymers for improved oil recovery: a literature review. J Pet Sci Eng 1998;19(3-4):265-80. URL: http://www.sciencedirect.com/science/article/pii/S092041059700048X. doi:10.1016/S0920-4105(97)00048-X.

[47]   Evani S. Enhanced oil recovery process using a hydrophobic associative composition containing a hydrophilic/hydrophobic polymer. US Patent 4814096, assigned to The Dow Chemical Company (Midland, MI); 1989. URL: http://www.freepatentsonline. com/4814096.html.

[48]   Langlotz B, Reichenbach-Klinke R, Spindler C, Wenzke B. Method for oil recovery using hydrophobically associating polymers. WO Patent 2012069477 assigned to Basf Se; 2012. URL: https://www.google.at/patents/WO2012069477A1?cl=en.

[49]   Wellington SL. Stabilizing the viscosity of an aqueous solution of polysaccharide polymer. US Patent 4218327, assigned to Shell Oil Company (Houston, TX); 1980. URL: http://www.freepatentsonline.com/4218327.html.

[50]   Pich R, Jeronimo P. Method of continuous dissolution of polyacrylamide emulsions for enhanced oil recovery (EOR). US Patent 8383560, assigned to S.P.C.M. SA (Andrezieux Boutheon, FR); 2013. URL: http://www.freepatentsonline.com/8383560.html.

[51]   Vanderhoff JW, Wiley RM. Water-in-oil emulsion polymerization process for polymerizing water-soluble monomers. US Patent 3284393, assigned to Dow Chemical Co.; 1966. URL: http://www.google.com/patents/US3284393.

[52]   Anderson DR, Frisque AJ. Process for rapidly dissolving water-soluble polymers. US Patent 3624019, assigned to Nalco Chemical Company (Chicago, IL); 1971. URL: http://www.freepatentsonline.com/3624019.html.

[53]   Pinschmidt Jr RK, Vijayendran BR, Lai TW. Crosslinked vinylamine polymer in enhanced oil recovery. US Patent 5085787, assigned to Air Products and Chemicals, Inc. (Allentown, PA); 1992. URL: http://www.freepatentsonline.com/5085787. html.

[54]   Trahan DO. Method and composition to remove iron and iron sulfide compounds from pipeline networks. US Patent 8673834; 2014. URL: http://www. freepatentsonline.com/8673834.html.

[55]   Purinton Jr RJ, Mitchell S. Practical applications for gelled fluid pigging. Pipe Line Ind 1987;66(3):55-6.

[56]   Kennard MA, McNulty JG. Conventional pipeline-pigging technology: Pt.2: corrosion-inhibitor deposition using pigs. Pipes Pipelines Int 1992;37(4):14-20.

[57]   Messner SF. Cleaning of pipelines with gel pigs (csotavvezetek tisztitasa geles csoma-lacokkal). Koolaj Foldgaz 1991;24(7):219-22.

[58]   Lowther FE. Method for treating tubulars with a gelatin pig. US Patent 5215781, assigned to Atlantic Richfield Company (Los Angeles, CA); 1993. URL: http://www. freepatentsonline.com/5215781.html.

CHAPTER V

Additives for General Uses

1. WATER SOLUBLE POLYMERS

There are monographs and reviews on water soluble polymers for use in the petroleum industry [1–4].

The basic technology of hydraulic fluids for technological applications, however not stressed to petroleum applications, is described in detail [5].

2. CORROSION INHIBITORS

Aqueous compositions often contain dissolved or entrained air which increases the rate of corrosion and deterioration of metal surfaces in the drill string and associated equipment [6].

Corrosion inhibitors are frequently introduced into oil and gas fluids to aid in maintaining infrastructure integrity. Corrosion inhibitors are added to a wide array of systems and system components, such as [7]:

- Cooling systems,
- Refinery units,
- Pipelines,
- Steam generators,
- Oil or gas producing equipment, and
- Production water handling equipment.

These corrosion inhibitors are used to combat a large variety of types of corrosion. A common type of corrosion encountered in pumping of a fluid that contains corrosive agents is a flow-induced corrosion. The degree of corrosion depends on a multitude of factors. These factors include the corrosiveness of the fluid, pipeline metallurgy, shear rate, temperature, and pressure [7].

The injection of a proper corrosion inhibitor at the appropriate location and optimum dosage can be extremely effective in reducing the rates of corrosion.

Water-Based Chemicals and Technology for Drilling,
Completion, and Workover Fluids
http://dx.doi.org/10.1016/B978-0-12-802505-5.00005-6

Copyright © 2015 Elsevier Inc.
All rights reserved.

2.1 Distribution in Aqueous Phases

Corrosion inhibitors have particular attention due to their inherent design to partition into the aqueous phase. The environmental impact of a corrosion inhibitor is often defined by three criteria [8]:

1. Biodegradation,
2. Bioaccumulation, and
3. Toxicity.

All these three criteria have benchmarks that must be met for a chemical to be permitted for use, with different emphasis on each depending on which regulator controls the waters.

2.2 Deterioration of Inhibitor

In some cases, the performance of a corrosion inhibitor may deteriorate in time, particularly in systems that have a tendency to accumulate significant quantities of solids [7].

Solids can build up to form a layer up to several centimeters thick. Deposits of such hydrocarbonaceous materials and finely divided inorganic solids form on the inner surfaces of the lines. These deposits can include sand, clays, sulfur, naphthenic acid salts, corrosion byproducts, and biomass bound together with oil.

These particles may become coated with the corrosion inhibitor subsequently become coated with additional quantities of heavy hydrocarbonaceous material. Such a layer is often referred to as *schmoo* in the petroleum industry.

Schmoo is a solid or paste-like substance that adheres to almost any surface with which it comes in contact and is particularly difficult to remove. Whenever possible, pipelines known to have such deposited materials or that form pools of water at low spots are routinely pigged to remove the material.

In many cases, however, it may not be feasible to pig lines due to the construction configuration, variable pipeline diameter, or the lack of pig launchers and receivers. The material often accumulates on the bottom or around the circumference of the pipe. Often, even after maintenance pigging, schmoo still resides inside pits in metal surfaces.

Such situations may create a significant risk for increased corrosion. Schmoo can also accumulate to a thickness such that it flakes off the inner surfaces of the pipe and deposits in the lower portion of a well, the lower portion of a line or the like, and plugs the line or the formation in fluid communication with the pipe [7].

The physical barrier formed by such a layer may also retard the diffusion of the corrosion inhibitor to the walls. Often, these solids have a strong affinity to corrosion inhibitors and may significantly reduce the in situ inhibitor availability.

The composition of the matter within the solids may form an ideal environment to foster bacterial growth the metabolic byproducts of these reactions are frequently highly corrosive. Such a microbiologically influenced corrosion process has been recognized as a significant problem in the industry for many years.

Additional challenges are encountered in water injection systems when material carried in the water causes plugging of the sand-face downhole. Such plugging often leads to reduction in water injection efficiency and a consequent reduction of the oil produced [7].

2.3 Inhibitive Chemicals

Subsequently some chemicals that are used as corrosion inhibitors are detailed.

2.3.1 Oxygen Scavengers

Oxygen scavenger additives such as sodium dithionite and mixtures thereof have been used. Sodium dithionite is available in powder or liquid form but is difficult to handle and presents a fire hazard when exposed to the atmosphere. When used in powder form, the sodium dithionite is typically suspended in a liquid carrier, insulating the oxygen reactive materials from the atmosphere. When used in liquid form, the sodium dithionite must be transported under climate control.

Another method for providing an oxygen scavenger additive composition is done by mixing in situ a stream of an alkali metal borohydride with a stream of an alkali metal bisulfite.

Thus, sodium borohydride reacts with sodium bisulfite to give sodium dithionite, which in turn yields sulfinic acid, and then ultimately hydrogen as the oxygen scavenger, as shown in Eq. 4 [6].

$$\begin{aligned} NaBH_4 + 8NaHSO_3 &\rightarrow 4Na_2S_2O_4 + NaBO_2 + 6H_2O \\ Na_2S_2O_4 + 2H_2O &\rightarrow 2(-SO_2H)NaSO_3 + H_2 \end{aligned} \quad (4)$$

Alternatively, the oxygen scavenger additive may by composed from reducing agents such as sodium dithionite, thiourea dioxide, hydrazine, and iron reduction additives [6]. Further, chelating agents are used, such as ethylenediaminetetraacetic acid (EDTA), diethylenetriamine pentaacetic acid (DTPA), or sodium tripolyphosphate (STPP). Improved mud additives for the treatment of oil wells in the presence of downhole iron have been proposed. These minimize the hole enlargement when salt beds are encountered during the drilling process. Also the need for fresh water in the control of salt deposition within the well is reduced.

Sodium or potassium ferrocyanide and phosphonic acid or a phosphonic acid salt are used [9]. The composition is added to the recirculating drilling mud and lessens the salt erosion even in the case of wells with high concentrations of iron.

It has been found that the addition of sodium or potassium ferrocyanide to brine increases the concentration of sodium chloride therein. When a crystallization reaction of the sodium chloride takes place, the resultant salt crystals become more pyramidal, i.e., dendritic instead of cubical. This increased salt concentration in the brine renders it particularly useful for well treatments.

A preferred phosphonic acid compound is diethylenetriamine penta(methylene phosphonic acid) available under the trade name Dequest® 2066A. The structure is shown in Figure 51.

The corrosion of iron occurs by a mechanism shown in Eq. 5.

$$2HCl + Fe \rightarrow H_2 + FeCl_2 \qquad (5)$$

Also carbon dioxide and hydrogen sulfide corrode iron as shown below.

$$H_2S + Fe \rightarrow H_2 + FeS \qquad (6)$$

$$CO_2 + H_2O + Fe \rightarrow H_2 + FeCO_3 \qquad (7)$$

Acid corrosion inhibitors have been developed that are biodegradable, non–bioaccumulating and low in toxicity [10]. Acetophenone can react

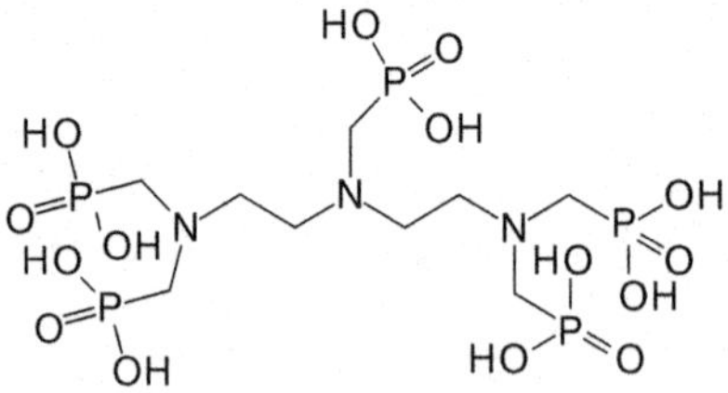

Figure 51 Dequest® 2066A.

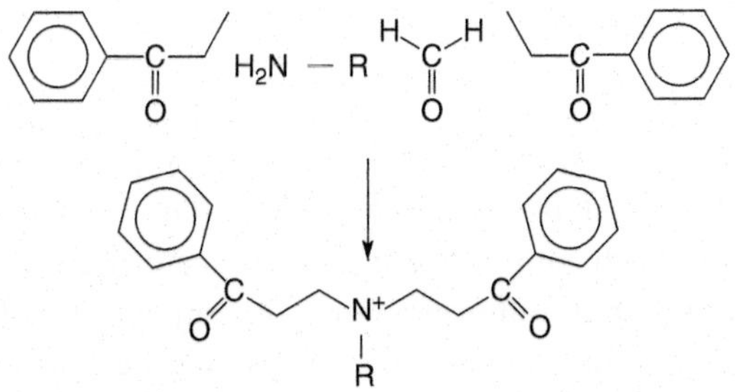

Figure 52 Mannich base inhibitor [10].

with an amine and formaldehyde to form a Mannich base inhibitor. The *Mannich* reaction was first described in 1912 [11]. The reaction is shown in Figure 52.

2.3.2 Quaternary Nitrogen Compounds

Quaternary nitrogen compounds have been used extensively as they form a film on the surface of steel, are stable over a wide range of pH and temperature, cost effective, efficient in sour conditions and inhibit microbially induced corrosion [8].

Also, imidazole based corrosion inhibitors have been appreciated [12–14]. However, due to their inherent biostatic properties their biotoxicity profile is often unacceptable and the compounds are not readily biodegradable [8]. This statement is in some contrast to an earlier publication [15]. Its gas been stated that it is possible to reduce the toxicity of a diamine or of an imidazoline by reacting these molecules with acrylic acids. The larger the number of acrylic acid molecules, the more the toxicity reduces [16, 17].

Quaternary nitrogen substituted pyridine compounds with long chain fatty esters have been claimed to have a lower environmental impact when compared to existing commercial treatments by virtue of their low toxicity, higher biodegradation, and lower bioaccumulation [8]. In addition, the corrosion inhibitors are less volatile, hence less malodorous than alkylpyridine corrosion inhibitors.

These compounds are prepared by the ring opening of epichlorohydrin to get 1-chloro-2,3-dihydroxy-propane. This compound is then condensed with a tall oil fatty acid to get a mixture of mono and bis esters. Then a substituted pyridine is added and finally the corresponding substituted pyridinium salt is formed [8].

3. BACTERIA CONTROL

3.1 Phosphonates

Phosphonates are maximally effective at high temperatures whereas sulfonated polymers are maximally effective at low temperatures [18]. Copolymers that contain both phosphonate and sulfonate moieties can produce and enhanced scale inhibition over a range of temperatures. A phosphonate end-capped vinylsulfonic acid/acrylic acid copolymer has been shown to be particularly useful in the scale inhibition of barium sulfate scale in water-based systems [18]. The basic issues of scale inhibitors are given in Table 60.

4. ANTIFREEZE AGENTS

An antifreeze is defined as an additive that, when added to a water-based fluid, will reduce the freezing point of the mixture [20]. Antifreezes are used in mechanical equipment in environments below the freezing point to prevent the freezing of heat transfer fluids. Another field of application is in cementing jobs to allow operation below the freezing point.

Hydrate control is not included in this chapter, but is discussed in another chapter, because of the relative importance and difference in chemical mechanism. Many chemicals added to water will result in a depression of the freezing point. The practical application is restricted, however, because of some other unwanted effects, such as corrosion, destruction of rubber sealings in engine parts, or economic aspects.

Table 60 Types of scale inhibitors [19]

Inhibitor type	Limitations
Inorganic poly(phosphate)s	Suffer hydrolysis and can precipitate as calcium phosphates because of temperature, pH, solution quality, concentration, phosphate type and the presence of some enzymes
Organic poly(phosphate)s	Suffer hydrolysis with temperature. Not effective at high calcium concentrations. Must be applied in high doses
Polymers based on carboxylic acids	Limited calcium tolerance (2000 ppm) although some can work at concentrations higher than 5000 ppm. Larger concentrations are needed
Ethylene diamine tetraacetic acid	Expensive

4.1 Theory of Action

Freezing point depression follows the colligative laws of thermodynamics at low concentrations added to water. At the same time, the boiling point generally will be increased. The freezing point depression can be readily explained from the theory of phase equilibria in thermodynamics.

In equilibrium the chemical potential must be equal in coexisting phases. The assumption is that the solid phase must consist of one component, water, whereas the liquid phase will be a mixture of water and salt. So the chemical potential for water in the solid phase μ_s is the chemical potential of the pure substance. However, in the liquid phase the water is diluted with the salt. Therefore the chemical potential of the water in liquid state must be corrected. x refers to the mole fraction of the solute, that is, salt or an organic substance. The equation is valid for small amounts of salt or additives in general:

$$\mu_s = \mu_l + RT \ln(1 - x). \tag{8}$$

The equation is best expressed in the following form:

$$\frac{\mu_s - \mu_l}{RT} = \ln(1 - x) \cong -x. \tag{9}$$

The derivative with respect to temperature will give the dependence of equilibrium concentration on temperature itself:

$$\frac{d \frac{\mu_s - \mu_l}{RT}}{dT} = -\frac{dx}{dT} = \frac{\Delta H}{RT^2}. \tag{10}$$

ΔH is the heat of melting of water. Because the heat of melting is always positive, an increase of solute will result in a depression of the freezing point. For small freezing point depressions, the temperature on the right hand side of the equation is treated as a constant. Furthermore, it is seen that additives with small molecular weight will be more effective in depressing the freezing point. Once more it should be noted that the preceding equation is valid only for small amounts of additive. Higher amounts of additive require modifications of the equation. In particular, the concept of activity coefficient has to be introduced. The phase diagram over a broader range of concentration can be explained by this concept.

4.2 Antifreeze Chemicals

Some data concerning the activity of antifreeze chemicals are presented in Table 61. Inspection of Table 61 shows that there are two different

Table 61 Antifreeze chemicals

Component	Concentration in water [%]	Depression of freezing point [°C]
Calcium chloride	32	−50.0
Ethanol	50	−38.0
Ethylene glycol	50	−36.0
Glycerol	50	−22.0
Methanol	50	−50.0
Potassium chloride	13	−6.5
Propylene glycol	50	−32.0
Seawater (6% salt)	–	−3.0
Sodium chloride	23	−21.0
Sucrose	42	−5.0
Urea	44	−18.0

Table 62 Depression of the freezing point in a mixture of ethylene glycol-water

Amount of ethylene glycol [%]	Depression of freezing point [°C]
10	−4
20	−9
30	−15
40	−24
50	−36

types of antifreeze chemicals, that is, liquids that are miscible over the full range of concentration with water and salts, often salts which are soluble only to a certain amount. In the case of liquids, a mixture of 50% with water is given. In the case of solids, the EG forms with water an eutectic point between 65% and 80% at around −70 °C. Pure EG will solidify at −14 °C, however. Mixtures of propylene glycol with water can supercool at higher concentrations of propylene glycol. So, the equilibrium freezing points cannot be measured.

The depression of the freezing point in a mixture of EG and water is shown in Table 62. The phase diagram of the binary system EG-water is plotted in Figure 53. Some organic antifreeze agents are depicted in Figure 54.

4.3 Heat Transfer Liquids

The classic antifreeze agents in heat transfer liquids are brine solutions and alcohols.

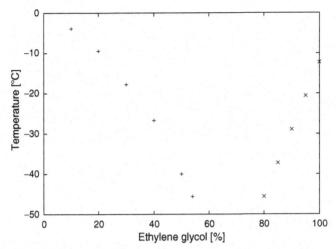

Figure 53 Phase diagram of the binary system for ethylene glycol-water.

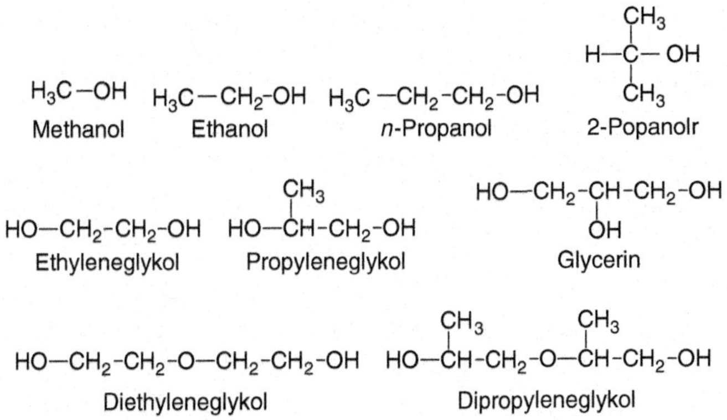

H₃C—OH   H₃C—CH₂-OH   H₃C —CH₂-CH₂-OH

$$H_3C-OH$$ $$H_3C-CH_2-OH$$ $$H_3C-CH_2-CH_2-OH$$

Methanol   Ethanol   n-Propanol

$$\begin{array}{c} CH_3 \\ | \\ H-C-OH \\ | \\ CH_3 \end{array}$$

2-Popanolr

$$HO-CH_2-CH_2-OH$$ $$\begin{array}{c} CH_3 \\ | \\ HO-CH-CH_2-OH \end{array}$$

Ethyleneglykol   Propyleneglykol

$$\begin{array}{c} HO-CH_2-CH-CH_2-OH \\ | \\ OH \end{array}$$

Glycerin

$$HO-CH_2-CH_2-O-CH_2-CH_2-OH$$

Diethyleneglykol

$$\begin{array}{c} CH_3 \qquad CH_3 \\ | \qquad | \\ HO-CH-CH_2-O-CH-CH_2-OH \end{array}$$

Dipropyleneglykol

Figure 54 Organic antifreeze agents.

4.3.1 Brines

Of the commonly used antifreeze agents, brines are the most corrosive to metals of the engines and exhibit scale deposition characteristics that are highly restrictive to heat transfer. Today brines (seawater) still find applications in offshore uses because they are cheap.

4.3.2 Alcohols

Alcohols, such as methanol and ethanol are readily available and are occasionally used despite significant disadvantages, such as low boiling points. During summer months significant amounts of alcohol can be lost due to evaporation. Such losses lead to a costly replacement of the additive. Furthermore, alcohols have low flash points, which may cause safety problems. Moreover, methanol is highly poisonous. Therefore, the use of alcohols has ceased almost completely in recent years.

4.3.3 Glycols

EG is not as active in depression of the freezing point as methanol, but it has a very low vapor pressure. Evaporation loss in a coolant system is due more to the evaporation of water than to the evaporation of EG. Furthermore, the flammability problem is literally eliminated. 1:1 mixtures of EG and water do not exhibit a flash point at all. EG-based antifreeze formulations may contain small amounts of other glycols such as diethylene glycol or triethylene glycol. Propylene-based glycols such as propylene glycol and propylene glycol ethers have limited use, especially in areas in which regulations about human toxicity apply. EG proves most effective in depression of the freezing point and heat transfer activities.

Properties of Glycol-Based Antifreeze Formulations

Pour Point   The desired concentration of an antifreeze agent will be governed by several features. The freezing point of a mixture is the point at which the first ice crystal can be observed. This does not mean, however, that this temperature would be the lowest allowable temperature in the application. In the case of heat transfer agents, the fluid will not function efficiently, but because the fluid will not freeze completely to a solid state it may still be operational. Pure water will expand by complete freezing at about 9%. The addition of antifreeze, such as EG, will significantly lower the amount of expansion, thus protecting the system from damage. At the freezing temperature the crystals are mainly water themselves. Therefore, the concentration of the antifreeze agent still in the solution will be increased. This causes a further depression of the freezing point of the residual liquid. At higher glycol concentrations the fluid never solidifies completely. The fluid becomes thick and taffy like. The point at which the fluid ceases to flow is referred to as the pour point. The pour point is significantly lower than the freezing point. However, the use of such a system down to the pour point will significantly increase the energy required for pumping. Furthermore, because of the decreased ability for heat transfer, it is generally not recommended to regularly use systems beyond the freezing point of the mixture.

Corrosion   Alcohols may be corrosive to some aluminum alloys. In an aqueous mixture corrosion may still occur because of dissolved ions from residual salts. At high temperatures and in the presence of residual oxygen from the air, glycols are oxidized slowly to the corresponding acids. These acids can corrode metals.

The inhibition of acid corrosion can be achieved by adding buffer systems that essentially keep the pH constant and neutralize the acids. For example, a formulation of 100 kg EG with 400 g KH_2PO_4, 475 g Na_2HPO_4, and 4 l of water is used as an antifreezing agent, which can be diluted accordingly with water, approximately 50:50. This formulation will be highly anticorrosive. Also, borax can be used to protect metal surfaces from corrosion.

Besides of pure chemical corrosion, solid products of corrosion in the system will give rise to erosive corrosion, in which the particles moving with the fluid will impact onto the surfaces and can remove protective surface layers. Such corrosion effects are most pronounced in regions of high fluid-stream velocity.

The most common corrosion inhibitors, which may form protective films on the metal surfaces, are borates, molybdates, nitrates, nitrites, phosphates, silicates, amines, triazoles, and thiazioles, e.g., monoethanol amine, urotropin, thiodiglycol, and mercaptobenzothiazole. The addition of such inhibitors does not effectively protect against corrosion [21]. Some corrosion inhibitors are shown in Figure 55.

Dibasic salts of dicyclopentadiene dicarboxylic acid are claimed to be active as corrosion inhibitors [22]. Certain salts of fatty acids (metal soaps), together with benzotriazole, are claimed to give synergistic effects for corrosion in antifreeze agent formulations [23].

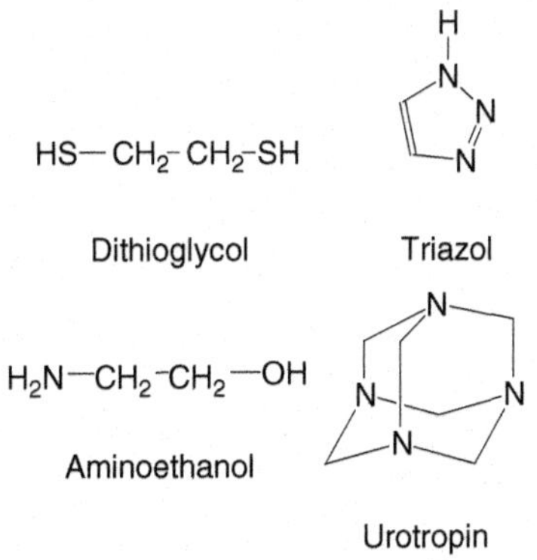

Figure 55 Corrosion inhibitors.

The choice of a corrosion inhibitor as an additive in antifreezing agents is also dependent on the mode of operation. For instance, cars are operated intermittently. Here the corrosion inhibitors must also protect the system when it is idle. Film forming silicates can protect the system while idle. This is especially true of aluminum parts, which are introduced in cars for the sake of weight reduction. But silicones can react with EG to form crosslinked polymers. These gels may clog lines.

The engines in the oil industry are usually heavy stationary diesels that run continuously. Also, aluminum is normally not used in this type of engine. For these types of engines, corrosion inhibitors for glycol systems based on silicate-forming films are not recommended, because of gel formation. Appropriate blends of corrosion inhibitors added to the glycol-water mixture to minimize corrosion problems in applications for coolants have been developed [24].

Therefore, coolant formulations for engines involved in natural gas transmission consist of phosphate for ferrous metal protection and a triazole for the protection of brass parts. Corrosion is discussed in detail in another chapter.

Foam Inhibitors　Although glycol-water formulations are not prone to foaming, mechanical and chemical factors may cause foaming in the system. The use of corrosion inhibitors and the presence of contaminants may enhance the tendency to form foams. For these reasons, antifoaming agents, such as silicones, polyglycols, or oils, are sometimes added.

Damage of Elastomers　Some elastomer sealings that are in contact with the antifreeze mixture may not be stable in such a medium because of consequences such as swelling. The compatibility of EG with certain plastics is shown in Table 63.

Table 63 Compatibility of ethylene glycol with some elastomers

Material	25 °C	80 °C	160 °C
Poly(urethane)	Good	Poor	Poor
Acrylonitrile-butadiene copolymer	Good	Good	
Styrene-butadiene copolymer	Good	Fair	Poor
Ethylene-propylene-diene copolymer	Good	Good	Good
Natural rubber	Good	Poor	Poor
Silicone rubber poly(dimethylsiloxane)	Good	Good	
Vinylidene fluoride-hexafluoropropene rubber	Good	Good	Poor

4.3.4 Biodiesel Byproducts

Glycol compounds likewise have environmental drawbacks and can be detrimental to aquatic life and to sewage treatment processes [25]. Other prior art deicing fluids, such as alcohols, have toxic effects and high volatility particularly in the low molecular weight range. Further, some of these may be the cause of offensive smell and fire danger. Furthermore, mono and polyhydric alcohols oxidize in the presence of atmospheric oxygen to form acids, which can increase corrosion of materials.

The production of biodiesel is of growing importance in order to reduce the dependence on fossil fuels. Triglycerides, the principal components of animal fats and of vegetable oils, are esters of glycerol with fatty acids of varying molecular weight.

There are three basic routes to production of biodiesel employing homogeneous systems [25]:

1. Base catalyzed transesterification of the oil,
2. Acid catalyzed transesterification of the oil,
3. Conversion of the oil to fatty acids by hydrolysis and subsequent esterification to biodiesel.

The base catalyzed route is the most popular because of the reaction efficiency, mild operating conditions and it requires only simple materials of construction.

Once the glycerol and biodiesel phases have been separated, the excess alcohol is removed via a flash evaporation process or by vacuum distillation. There, glycerol remains as byproduct. One impediment in this field is to find a profitable use for the glycerol containing byproduct of the conversion reaction.

Deicing compositions comprising byproducts of the reaction of triglycerides to produce monoesters for biodiesel have been developed. It has been found that excellent deicing compositions can be obtained from the byproduct of reactions with triglycerides.

The raw glycerol can be combined with other deicing compounds. The reduction in freezing point dependent on the compositions is shown in Table 64.

5. THICKENERS

5.1 Dendritic Polymers

A dendritic comb-shaped polymer can be used a thickening agent. The preparation and application has been described [26]. Monomers for dendritic polymers are shown in Figure 56.

Table 64 Freezing point depression [25]

Compound	Amount [%]							
Glycerol	50	25		25			25	
Potassium carbonate		25	47					
Potassium acetate				25	25	50		
Sodium lactate							25	50
Water	50	50	53	50	75	50	50	50
Freezing point [°C]	−23	−37	−20	−41	−18	−60	−39	−32

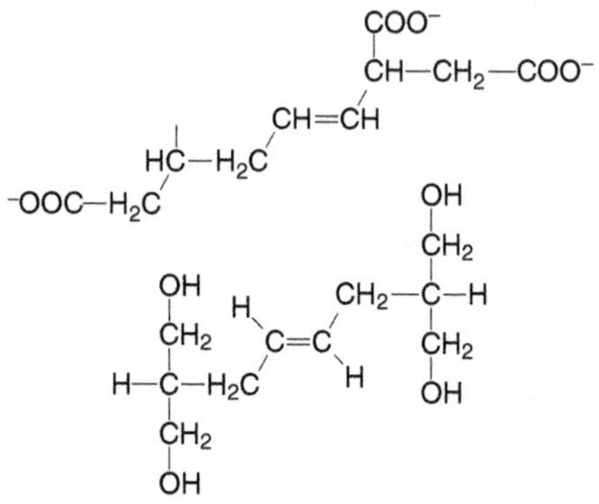

Figure 56 Monomers for dendritic polymers [26].

These monomers are copolymerized with other water soluble unsaturated comonomers, such as acrylamide, vinylpyrrolidone monomers, or 2-acrylamido-2-methyl-1-propane based monomers.

These polymer types can used in polymer flooding for class II oil reservoir in oilfields. They can be quickly dissolved in produced water and the viscosity of the polymer solution more than 30% higher than those of common poly(acrylamide) with a comparable molecular weight [26].

5.2 Ammonium Salts

Alkyl amido quaternary ammonium salts, and formulations thereof, can be used as thickening agents in aqueous-based fluids [27]. Examples are erucyl amidopropyltrimethyl ammonium quaternary salt, dimethylamidopropyltrimethyl ammonium chloride, and dimethylalkylglycerolammonium chloride.

The quaternary ammonium salts exhibit an improved gelling characteristics. The compositions may find use as gellants in fluids used in oil recovery operations such as fracturing fluids, completion fluids, and drilling fluids.

For example, the thickened fluids are capable of suspending proppant particles and carrying them to a fracture site. The gelled fluids also diminish the loss of fluid into a fracture, thereby improving the efficiency in a fracturing process [27].

5.3 Organoclay Thickeners

Hectorite compounds, in addition to many other swellable clays, have been proposed as a thickener for well-servicing fluids. These layer silicates exhibit an excellent thermal and electrolyte stability [28]. In water or electrolyte solutions, these swellable layer silicates form clear gels with excellent thickening and thixotropic properties. These properties are not substantially adversely affected by electrolyte loading or temperature stressing. The silicates can be used in conjunction with synthetic polymers.

Also, organoclays may serve as thixotropic agents in water-based systems. A preferred process for the preparation involves the following steps [29]:

1. Crushing run-of-mine clay to less than 100 mesh and gently dispersing the raw clay in a polymer-rich phase of an aqueous biphasic extraction system at a solids concentration sufficient to provide a smectite clay concentration of 2-6% in the polymer-rich phase of the aqueous biphasic extraction system.
2. Contacting the polymer-rich phase containing dispersed clay ore with an aqueous salt solution such as sodium carbonate, sodium sulfate, or sodium phosphate. Treatment with the salt solution results in sodium being added at ion exchanges. The components are mixed gently at room temperature. Elevated temperatures may also be used to speed clay dispersion, water swelling, and ion exchange processes.
3. Separating liquid phases from each other, whereby mineral impurities are carried with a salt-rich liquid phase, while the sodium-exchanged organoclay is carried with the polymer-rich liquid phase. The resultant clay also includes a level of low-molecular-weight, water-miscible polymers such as poly(ethylene glycol), poly(vinylpyrrolidone), and poly(vinyl alcohol), adsorbed from polymers present in the polymer-rich phase.
4. Recovering the organoclay from the polymer-rich liquid phase by flocculation with an appropriate, high-molecular-weight polymer. Types of polymers appropriate for the flocculation of the organoclay are those normally used in mineral processing for solid/liquid separation of clay systems and can be non-ionic, cationic, or anionic polymers with molecular weights of 1-5 MDa.

5. Washing the organoclay to remove excess polymer and drying at a temperature low enough to avoid excessive oxidation of a polymeric surface coating on the clay. Avoidance of excessive oxidation typically requires drying temperatures at or below 100 °C. Drying time can be reduced significantly by further modifying surface properties of the organoclay with low-molecular-weight polymers that increase hydrophobicity of the clay. Examples of such polymers are poly(propylene glycol), copolymers from ethylene glycol and propylene glycol, or ethylene glycol, propylene glycol, and butylene glycol. The organoclay surface can also be modified to improve drying by a partial conversion of the ion exchange sites with onium ions that are bearing hydrophobic alkyl groups.

An advantage of the above described process is that the clays produced through the aqueous biphasic extraction system possess a significantly smaller average particle size than those produced through traditional mineral processing techniques involving screens, hydrocyclones, and centrifuges.

This significantly smaller average particle size occurs by virtue of the fact that the polymers used in generating the aqueous biphasic extraction system also aid in dispersing the clay using only gentle stirring, in contrast to current approaches which subject the clay slurry to high shear in order to disperse the clay and thereby reduce the average particle size of the clay.

The process produces organoclay slurries with a small average particle size, for example 0.196 μm, using only gentle stirring with a mechanical mixer at room temperature. The gentle stirring enables processing equipment to maintain a relatively long life span compared to highly energy intensive processes that lead to rapid abrasion of processing equipment [29].

6. SURFACTANTS

An emulsion-based cleaning composition for oilfield applications has been described [30].

Previously used surfactants have been ethoxylated sorbitan fatty acid esters. Agents based on ethoxylated sorbitan fatty acid esters exhibit a good cleaning effect, but with regard to biodegradability and toxicity, such agents cannot yet fulfill all requirements that increasingly stringent environmental legislation requires.

The use of a mixture of an alkyl polyglycoside and a monoester of glycerol and a fatty acid, in particular, glycerol and oleic acid, is preferred for these reasons.

7. LUBRICANTS

7.1 Synergism

Aqueous mixtures of nonionic monoglyceride surfactants and xanthan gum were investigated with in order to correlate their lubricity and solubility in water, as well as using surface tension and contact angle measurements [31]. The results revealed that monoglycerides behave as excellent lubricants in water, with a steady decrease of the friction coefficient as the chain length of the hydrocarbon moiety increases.

The monoglycerides can reduce the friction coefficient even further when used in xanthan gum suspensions. This finding suggests that they are probably forming a complex with the polysaccharide that shows a synergy toward their performance as lubricants.

Adsorption experiments onto iron oxide nanoparticles also yielded an evidence of the interaction between these molecules, which favors their adsorption on the metal surface [31].

7.2 Nano-Attapulgite

Nano-attapulgite has been used to improve the tribological properties of drilling fluids [32].

Attapulgite was purified, synthesized, characterized, functionalized, and tested in the form of nano particles with 10-25 nm diameter for its effectiveness in order to tailor the rheology of drilling fluids swiftly to reduce friction, discourage use of other expensive additives, and to improve their functionality. Nano-attapulgite modified drilling fluids change mainly the coefficient of friction between the drillstring and the wellbore [32].

8. FOAMS

Aerated liquids and foams are frequently employed as wellbore fluids during drilling, completion, workover, and production operations [33]. The term aerated refers to the existence of distinct gas and liquid phases. Aerated liquids are distinguishable from foams by the absence or the presence of small amounts of surface active agents that promote the gas–liquid interphase dispersibility.

Aerated liquids and foams are particularly useful when reduced hydrostatic pressures in wellbores are desired. The reduced hydrostatic pressure or head is made possible by the lower bulk densities of aerated liquids and foams when compared to their liquid counterparts. Reduced hydrostatic heads are favored in situations where the wellbore fluids are exposed to under-pressured geological strata and where a minimum invasion of the strata by the wellbore fluid is desired, further during the perforation of a geological strata at under-balanced conditions. This means that the wellbore pressure is less than fluid pressure. There, the reduced head helps to minimize plugging of the perforation caused by the movement of fines into the perforation from the wellbore [33]. Aerated liquids are comprised of a gas and a liquid and may additionally contain additives such as corrosion inhibitors and suspended solids such as mud and sand.

Foams are highly dispersed gas in liquid two-phase systems. A foam is physically composed of gas bubbles surrounded by a surfactant-bearing aqueous film and nominally contains a gas, a surfactant which functions as a foaming agent, and water. The quality of the foam or foam quality is defined to be the volume percent of gas in the two-phase mixture. The water used in foams may be chemically pure, fresh, or contain varying degrees of salinity or hardness. The foam may additionally contain other additives which alter the properties of the foam. Such additives include selected polymers or bentonite, which increase film strength, thereby forming stiff or gelled foams and corrosion inhibitors which function to decrease foam corrosivity in metal-bearing systems.

The rheological properties of foam are complex and provide a separate and distinct basis for distinguishing foam from aerated and non-aerated liquids. The theological properties are dependent upon many parameters such as bubble size and distribution, fluid viscosity, foam quality, and the type of foaming agent. The use of a foam in place of conventional fluids generally reduces the degree of fluid invasion from the wellbore into the surrounding strata [33].

Furthermore, the foam which does invade the strata generally contains a high volume percentage of gas which upon pressurization possesses significant energy. Upon depressurization, the stored energy which is released causes a significant portion of this fluid to be returned to the wellbore. In comparison to conventional fluids, a foam also possesses at low linear flow velocities excellent carrying capacities, particularly for water and solids. These properties are particularly useful when conducting drilling, completion, and workover operations on vertical and horizontal wells and even more so when low pressure, semi-depleted or water-sensitive formations are encountered.

The unique properties of foam also enable the use of coiled tubing units during workover operations. The use of such units results in significant savings of time and money because downhole operations can be performed without removal of the wellbore tubing [33].

8.1 Foaming Agents

Foaming agents are summarized in Table 65 and in Figure 57.

Cocoamidopropylbetaine is also addressed as 2-[(3-dodecanamidopropyl) dimethylaminio]acetate. Betaines are multipurpose foaming agents [34]. Betaines are offering excellent detergency and interfacial tension-reducing properties. Betaines are particularly applicable to a water-sensitive reservoir, i.e., a reservoir, which is slow to return water-based treatment fluids. Such reservoirs are often characterized by low permeability, and moderate amounts of clay or shale. Betaines offer superior performance in high-temperature wells, where other foamers degrade.

8.2 Foamed Cement Slurries

Foamed cement slurries have included foaming and stabilizing additives which include components such as isopropyl alcohol that interfere with aquatic life. In addition, one or more of the components are often flammable and make the shipment of the foaming and stabilizing additives expensive.

Foamed hydraulic cement slurries which include environmentally benign foaming and stabilizing additives that do not include flammable components have been developed [35].

Table 65 Foaming agents [33]

Compound	Compound
N-Acrylsarcosinate	Sodium N-acryl-N-alkyltaurate
Alkyl sulfate	Ethoxylated and sulfated alcohol
Ethoxylated and sulfated alkyl phenol	Fatty acid diethanolamide
Amine oxide	Alkyl betaine
Amidopropylbetaine	Cocoamidopropylbetaine

N-Acrylsarcosinate   Glycine betaine

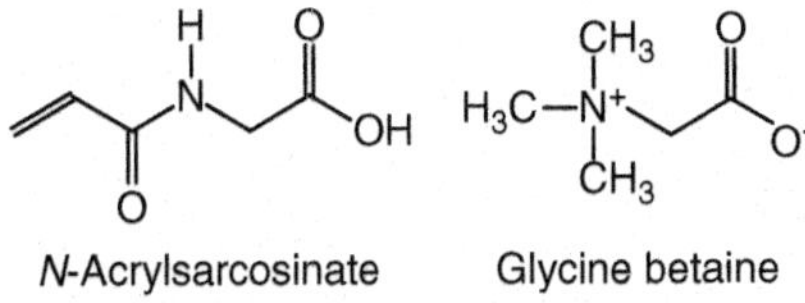

Figure 57 Betaine based foaming agents.

An environmentally benign additive for foaming and stabilizing a cement slurry is a mixture of an ammonium salt of an alkyl ether sulfate surfactant, a cocoamidopropyl hydroxysultaine surfactant, a cocoamidopropyl dimethylamine oxide surfactant, sodium chloride, and water [35].

8.2.1 Defoaming Agents

Hydraulic cement compositions often comprise defoamers. Defoamers are utilized, as components in well treatment fluids to prevent the formation of foam or the entrainment of gas during the preparation and placement of the well treatment fluid in the subterranean formation [36].

Defoamers also have been utilized for breaking foamed fluids. For example, when an operator desires to dispose of a foamed well treatment fluid aboveground, the operator may add a defoamer to the well treatment fluid to break the foam, and thereby facilitate disposal of the well treatment fluid. Defoamers often are included in cement compositions, in order to ensure a proper mixing and to provide adequate control over the density of the cement composition.

A variety of defoamers are known. However, some of them may have undesirable environmental characteristics or may be limited by strict environmental regulations in certain areas of the world.

Cement compositions with environmentally compatible defoamers have been described [36]. Such environmentally compatible defoamers comprise octanol, hexanol, or butanol and a surfactant.

9. GELLING IN WATER-BASED SYSTEMS

9.1 Xanthan Gum

The in situ gelation of aqueous solutions of xanthan gum can used to treat oil spills in soil as a first aid method. In experiments, the gelling reaction has been carried out using both Cr^{3+} and Al^{3+} cations. Cr^{3+} takes around 1 h to gel, whereas Al^{3+} gels xanthan at low pH almost instantaneously. Aqueous solutions of xanthan exhibit a shear thinning behavior, which is highly favorable for these applications [37].

9.2 Carboxymethyl Cellulose

A mixture of lignosulfonate with modified CMC and metal ions as crosslinking agents has been suggested as a plugging agent [38]. CMC, modified with poly(oxyethylene glycol) ethers of higher fatty alcohols, combines the properties of a surfactant and CMC.

Thus it reduces the viscosity of the composition and increases the strength of the produced gel. Sodium and potassium bichromates act as crosslinking agents. Ionic crosslinks are formed as a result of the reaction of Cr^{3+} and Ca^{2+} ions with molecules of modified CMC.

A gel-forming composition is obtained by mixing aqueous solutions of the respective components. Highly mineralized water also can be used and the gelation time can be controlled by changing the contents of $CaCl_2$ and bichromates.

9.2.1 Poly(dimethyl diallyl ammonium chloride)

Poly(dimethyl diallyl ammonium chloride) is a strongly basic cation-active polymer. A mixture of poly(dimethyl diallyl ammonium chloride) and the sodium salt of CMC, which is an anion-active polymer, is applied in an equimolar ratio [39] in aqueous sodium chloride solution. The proposed plugging composition has high efficiency within a wide pH range.

9.2.2 Lignosulfonate and Carboxymethyl Cellulose

An aqueous solution of 3-6% lignosulfonate and 2-8% CMC, modified with poly(oxyethylene glycol) ethers of higher fatty alcohols form the base of a plugging system [38]. Lignosulfonate is a waste product from the cellulose-paper industry. Furthermore, as crosslinking agents, sodium or potassium bichromate and calcium chloride are added in amounts from 2% to 5%. The composition is obtained by mixing aqueous solutions of the components in a cement mixer.

A mixture of polymers that can serve as a plugging solution when taken in an equimolar ratio consists of poly(dimethyl diallyl ammonium chloride), which is a strongly basic cation-active polymer, and the sodium salt of CMC, which is an anion-active polymer. The aqueous solution contains 0.5-4% of each polymer. Gelling occurs because the macro ions link together from different molecules. The proposed plugging composition has high efficiency within a wide pH range [39].

9.3 Poly(acrylamide)-Based Formulations

Aqueous solutions of PAM may be used as plugging solutions for high-permeability formations. Partially hydrolyzed poly(acrylamide) (PHPA) also has been used [40] and completely HPAN has been proposed [41].

The polymer solutions are pumpable as such, but in the presence of multivalent metal ions, gels are formed. The gel formation is caused by an intermolecular crosslinking, in which the metal ion forms bonds to the polymer.

The metal ions are often added as salts of organic compounds, which form chelates. This causes a delayed gelation. Likewise, the components of the gelling agent are pumped down in two stages. Some metal cations cannot be used with brines. On the other hand, brines are often produced from wells, and it is desirable to find uses for them to avoid disposal processes.

9.3.1 Delayed Gelation
Complexing Agents
Delayed gelation can be achieved by adding complexing agents to the mixture. The metal ions are initially complexed. Therefore, all components of the gelling composition can be injected simultaneously. It is possible to dissolve the mixture in produced brines that have high salinity. The use of produced brines eliminates the need to treat or dispose of the brines.

Examples of the components of the gelling composition are a water soluble polymer such as PAM, an iron compound such as ferric acetylacetonate or ammonium ferric oxalate, and a ketone such as 2,4-pentanedione [42]. The composition forms a temporary gel that is useful for the temporary plugging of a formation. The temporary gels that are formed will disappear after 6 months. Such complexes are shown in Figure 58.

Adjustment of pH
Some organic reagents hydrolyze in aqueous solution at elevated temperature and release ammonia. Examples are urotropin and urea. The hydrolysis of urea is shown in Figure 59.

Urotropin yields by hydrolysis formaldehyde and ammonia. In this way, the pH is increased. The chemical reactions necessary for the formation of a gel with other components of the mixture can then take place.

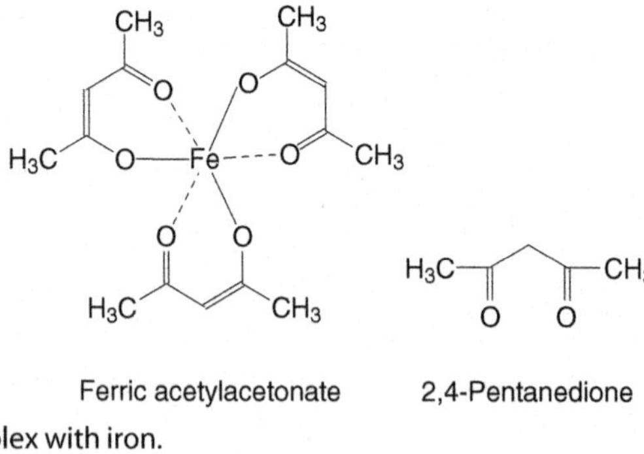

Ferric acetylacetonate   2,4-Pentanedione

Figure 58 Complex with iron.

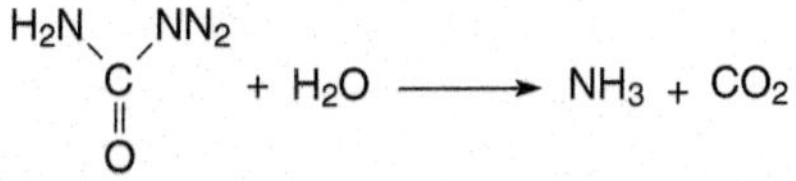

$$\underset{\substack{\| \\ O}}{\overset{\substack{H_2N \diagdown \diagup NN_2 \\ C}}{}} + H_2O \longrightarrow NH_3 + CO_2$$

Figure 59 Hydrolysis of urea with water.

9.3.2 Poly(acrylamide) and Urotropin-Based Mixture

In Table 66 a recipe for a PAM-based mixture is shown [43]. The gel-forming properties of such a PAM and urotropin-based mixture are shown in Table 67.

9.3.3 Reinforcement by Fibers

Fibers can be added to a gelation solution [44, 45]. Fibers that will not interfere with the gelation process and will provide adequate reinforcement must be chosen. In addition, they should not adversely affect the ability of the solution to be pumped and injected. In particular, glass fibers and cellulosic fibers meet the requirements as reinforcing fibers for plugging solutions.

9.3.4 Metal Ions and Salts as Crosslinking Agents

Waste Materials

Waste materials from other processes have been established to be active, such as waste from galvanizing processes [46]. In this case the components must be placed in two portions.

Table 66 Gel-forming composition based on poly(acrylamide) [43]

Material	Amount [%]
	0.05 to 3
Poly(acrylamide)	
Urotropin	0.01 to 10
Sodium bichromate	0.01 to 1
Water	100

Table 67 Gel-forming time at various temperatures [43]

Temperature [°C]	Time [h]
	10 to 18
60	
80	6 to 22
120	4.5 to 7

Iron and chromic salts from lignosulfonate are also a source for metal ions [47]. Lignosulfonates are waste products from the paper industry.

Chromium (III) Propionate

A chromium (III) propionate-polymer system is suitable for gelation treatments in oil fields, where fresh water is not available [48, 49]. It produces good and stable gels in hard brines, as well as in fresh waters. The effectiveness of treatment is a function of gelling agent concentration (i.e., the higher the concentration of chromium propionate, the higher the residual resistance factors).

The process can be used for in-depth treatment in the same way as with aluminum citrate and for near-well treatment. Produced brines that contain iron and barium can be used. In comparison with aluminum citrate, the use of chromium (III) propionate gives more effective in-depth treatment at only half of the crosslinker concentration. This makes the chromium propionate process even more attractive for freshwater applications.

Gelation Process and Gel Breaking

The gelation process of PAM with chromium ions takes place through coordination bonding with the nitrogen moiety [50]. Studies have indicated that better results can be expected in certain concentration ranges of the reactants. Gels formed at neutral pH have been observed to be comparatively stable and to sustain the reservoir temperature for 50 days. Solubility in chemicals like HCl, mud acid, and hydrogen peroxide indicates that these chemicals can be used as breakers.

Environmentally Safe Gel Breakers

Viscous well treating fluids are predominately water-based fluids containing a gelling agent comprised of a polysaccharide such as guar gum. However, some of the components in the treating fluids, in particular, delayed breakers are not environmentally benign [51].

Boron based crosslinking agents include boric acid, disodium octaborate tetrahydrate, sodium diborate, pentaborates, and minerals containing boron that release the boron upon hydrolysis. Poly(succinimide) and poly(aspartic acid) are delayed breakers that are environmentally benign. They chelate the boron and break the well treating fluid into a low viscosity fluid [51].

A gelled and crosslinked viscous treating fluid was prepared from fresh water, guar gum a caustic compound for adjusting the pH to alkaline, boric acid as the crosslinking compound, and poly(succinimide) as breaker. The

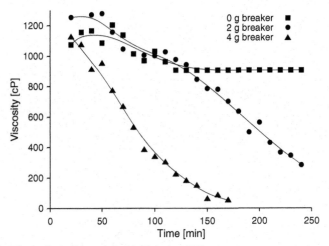

Figure 60 Effect of a poly(succinimide) breaker [51].

temperature of a sample of the treating fluid was raised to about 121 °C and the viscosity was measured over time using a Brookfield viscometer. The results of these tests are shown in Figure 60.

Sodium perborate tetrahydrate is an environmentally safe polymer breaker which is well suited for breaking viscous water-based fluids [52]. However, sodium perborate tetrahydrate has limited solubility in water. The solubility of sodium perborate tetrahydrate in fresh water or salt water can be increased by combining a solubility increasing boron complex or ester forming compound with the sodium perborate tetrahydrate. Complex or ester forming compounds are mannitol, mannose, galactose, glycerol, citric acid, tartaric acid, alkali poly(phosphate)s, phenols with attached polyhydroxyls, xanthan, guar, pectin, glycolipids, and glycoproteins.

9.3.5 Bis-quaternary Nitrogen Compounds

Bis-quaternary corrosion inhibitors and schmoo removers with low environmental impact and superior performance have been developed [7]. The synthesis of such bis-quaternary compounds is exemplarily shown in Figure 61.

Examples of suitable formulations containing bis-quaternary corrosion inhibitors are shown in Table 68.

The compositions can be used in any system exposed to fluids containing a metal corrosion agent where improved corrosion inhibition is desired. The compositions may be used in any component or any part of the oil and gas system where hydrocarbonaceous deposits are a concern [7].

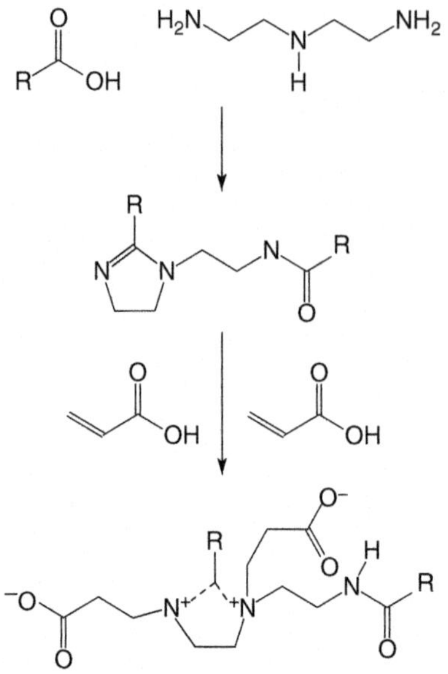

Figure 61 Synthesis of bis-quaternary compounds (R fatty acid residue) [7].

Table 68 Formulations with bis-quaternary corrosion inhibitors [7]

Compound	Amount [%]
Aromatic naphtha	0-75
Bis-quaternized compounds	5-50
Quaternary ammonium compound	0-20
Glycolic/thioglycolic acid	0-20
Acetic acid	0-20
Water or other solvent	0-95
Other optional components	0-95

For example, one technique in which such a formulation can be used is the squeeze treating technique, whereby the treating formulation is injected under pressure into a producing formation, adsorbed onto the strata, and desorbed as the fluids are produced. Also, the above described formulation may be used to enhance oil and gas production by optimizing water infectivity rates in water injection systems by removing hydrocarbonaceous deposits from such systems. Further, the formulation can be added in waterflooding operations of secondary oil recovery, and in well–acidizing operations [7].

Aluminum Citrate

Aluminum citrate can be used as a crosslinking agent for many polymers. The gels are made of low concentrations of polymer and aluminum citrate in water. This crosslinking agent provides a valuable tool, in particular, for in-depth blockage of high-permeability regions of rock in heterogeneous reservoirs. The formulations can be mixed as a homogeneous solution at the surface.

PHPAs, CMC, poly(saccharide)s, and acrylamido methylpropane sulfonate have been screened to investigate the performance of aluminum citrate as a chelate-type crosslinking agent. An overview of the performance of 18 different polymers has been presented in the literature [53]. The performance of the colloidal dispersion gels depends strongly on the type and the quality of the polymer used.

The gels were mixed with the polymers at two polymer concentrations, at three polymer-to-aluminum ratios, and in different concentrations of potassium chloride. The gels were quantitatively tested 1, 7, 14, and 28 days after preparation.

Interactions of Metal Salts with the Formation

Interactions of metal salts with the formation and distribution of the retained aluminum in a porous medium may significantly affect the location and strength of gels. This interaction was demonstrated with PAM-aluminum citrate gels [54]. Solutions were displaced in silica sand.

The major findings of this study are that as the aluminum to citrate ratio increases, the aluminum retention increases. Furthermore, the amount of aluminum retained by silica sand increases as the displacing rate decreases. The process is reversible, but the aluminum release rate is considerably slower than the retention rate.

The amount of aluminum released is influenced by the type and the pH level of the flowing solution. The citrate ions are retained by silica sand primarily as a part of the aluminum citrate complex. Iron, cations, and some divalent cations cannot be used in the brine environment.

9.3.6 Bentonite Clay and Poly(acrylamide)

A water-expandable material based on bentonite clay and PAM is added to the circulating drilling solution [55]. The material expands in water to 30-40 times its initial volume. The swelling takes place within 2-3 h. During the circulation of the drilling solution, the material enters the cracks and pit spaces of the natural stratal rock. When expanding slowly, the material converts within 30–40 min into the plugging material, which is strongly fixed to the rock.

Tests showed that the additive effectively prevents the absorption of the drilling solution on the stratal rock as a result of the production of a strongly adhering and insulating film, which is not dislodged even after subjecting the material to an excess pressure of 3 atm for 2 h. The time of expansion of the material is sufficiently slow to let it permeate into the slits and cracks of the stratal rock, yet is sufficiently quick to provide a compact insulation.

9.3.7 Thermal Insulation Compositions

Uncontrolled heat transfer from production tubing to the outer annulus, in particular, in deepwater riser sections can cause the deposition of sludge, paraffin, and asphaltene materials. Also the formation of gas hydrates may be enhanced and shut-in time for unplanned downtime or remedial operations may be limited. Deepwater risers can be insulated externally or insulated by placing nitrogen gas into the riser annulus [56]. Recently, a new water based thermal insulating fluid system has been developed and used in oil field applications. This system reduces the convection and provides a rheological profile to facilitate the fluid placement into the riser annulus.

A lightweight riser and insulating fluid has been developed. A special test facility was used, which featured a 20 m long full size aluminum riser with electrically heated production pipe, which was filled with the insulation fluid. Several formulations have been evaluated, both mineral oil and water-based fluids.

Undesired heat loss from production tubing as well as uncontrolled heat transfer to outer annuli can be detrimental to the mechanical integrity of outer annuli, because productivity losses from the well due to deposition of paraffin and asphaltene materials, accelerate the formation of gas hydrates, and destabilize the permafrost in arctic type regions [57].

Such fluids can be added either into the annulus or riser, and these fluids can effectively reduce the undesired heat loss. The composition contains a solvent of low thermal conductivity and a viscosifying polymer or a gelling agent.

The organic solvents include ethylene glycol, propylene glycol, glycerol, and diethylene glycol. The thermal conductivities of glycols are shown in Table 69.

The thermal conductivity of an aqueous DOWFROST® solution is shown in Figure 62.

Table 69 Thermal conductivities of alcohols and water types at room temperature [58]

Compound	Thermal conductivity [W (m K)$^{-1}$]
Isobutanol	0.134
Isopropanol	0.135
Propylene glycol	0.147
n-Pentanol	0.153
n-Butanol	0.167
n-Propanol	0.168
Diethylene glycol	0.203
Ethylene glycol	0.252
Glycerol	0.285
Ethylene glycol/water (75/25)	0.336
Ethylene glycol/water (50/50)	0.430
Freshwater	0.609
Seawater (7% salt)	0.7

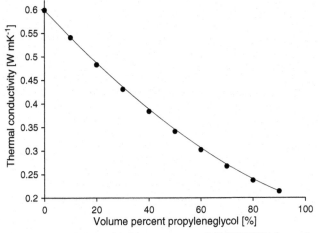

Figure 62 Thermal conductivity of an aqueous DOWFROST® (propylene glycol) solution [59].

The biochemical oxygen demand for propylene glycol approaches the theoretical oxygen demand value in a standard 20 d test period. This indicates that these materials are biodegradable and should not concentrate in common water systems. The possibility of spills in lakes or rivers, however, should be avoided, since rapid oxygen depletion may have harmful effects on aquatic organisms [59].

The solvent imparts low thermal conductivity to the composition and thereby provides highly desirable thermal insulation. The gelling agent is an AAm copolymer. The copolymers can be slightly crosslinked, e.g., with N,N'-methylenebisacrylamide. Other classes of polymers may be guar derivatives [57]. The crosslinking agents are based on borate compounds or zirconium or titanium complexes.

While water is preferably not used in conjunction with the solvent. However, water may be used in small amounts in the composition, such as a portion of a crosslinking system or the buffer system.

The thermal insulating composition can be prepared on the surface and is then pumped through the tubing in the wellbore or in the annulus. In particular, the fluid is a packer or riser fluid. Actually, the composition is acting in two modes [57]:

1. It serves to prevent heat transfer and buildup in the outer annuli and
2. It serves to retain the heat within the produced hydrocarbons.

9.4 Poly(acrylic acid)

Polymers of AA and methacrylic acid (MA) have been tested for their gel formation ability [60]. They are used with gel-forming additives similar to those described for PAMs. Also, mixtures of latex with methacrylate-MA copolymer as an additive have been described as plugging agents [61].

9.5 Alkali-Silicate Aminoplast Compositions

A composition to be injected into the ground for plugging consists of an alkaline metal silicate and an aminoplastic resin [62, 63]. Urea-formaldehyde, urea-glyoxal, or urea-glyoxal-formaldehyde condensation products are suitable. The composition has been suggested to be useful for a reservoir rock that is subjected to enhanced recovery methods. It can also be used for consolidation of ground and in building tunnels, dams, and other underground structures of this type.

9.6 Borates

Boric acid is a weak inorganic acid and the borate ion is not set free until the pH is sufficiently high to react with the more firmly bonded second and third hydrogen atoms [64].

The borate ion can form complexes with several compound types, e.g., polysaccharides like guar and locust bean gum as well as with poly(vinyl alcohol). At a high pH greater than 8, the borate ion exists as such and this ion can make crosslinking sites and thus cause gelling. In contrast, at a lower pH, the borate is associated firmly with the hydrogen and is not available for crosslinking reaction, thus the gelation caused by borate ions is reversible. When boric acid or borax is added to a 1% fully hydrated guar solution, the solution will gel.

It is well known that organic polyhydroxy compounds with hydroxyl groups positioned in the *cis*-form on 1,2 or 1,3 positions can react with borates to form complexes with five or six membered rings [65]. In the alkaline region these complexes form didiol complexes. The reaction is fully reversible with changes in pH. Therefore, an aqueous solution of such a polymer will gel in the presence of borate when the solution is made alkaline and will liquify again when the pH is lowered below about 8. Polymers that contain an appreciable content of *cis*-hydroxyl groups are guar gum, locust bean gum, dextrin, poly(vinyl alcohol).

Concentrated stable solutions of boric acid can be prepared by the in situ formation of an alkali metal or ammonium salt of an α-hydroxy carboxylic acid such as citric, lactic, and tartaric, in an aqueous, boric acid slurry [64]. Comparatively large quantities of boric acid or borax solids are slurred in an aqueous solution of the α-hydroxy carboxylic acid followed by the addition of sodium, potassium or ammonium hydroxide to achieve the in situ formation of the acid salt, until a pH of 6.5 is reached. This method of preparation results in a clear, stable solution of boric acid with nearly neutral pH. The thus prepared borate ion solution remains stable over an extended period of time and through temperature extremes and can be used with numerous guar solutions.

9.7 Reversible Gelling Under Shear

Water soluble polymers can be tailored to have gelling or viscosifying properties.

The most frequently used polymers are vegetable gums, which endow the solution with non-Newtonian properties with low viscosity at high shear rates. However, for certain applications, just the opposite behavior is desirable.

For example, a method for secondary recovery of hydrocarbons consists of injecting a flushing fluid, such as water, to which polymers are added in order to increase its viscosity to displace the hydrocarbons towards the production well. A fluid in which the viscosity increases reversibly with shear forces could minimize the viscous fingering and could render the flushing front uniform by preventing pockets of non flushed hydrocarbons from forming.

Compositions with rheo-viscosifying properties Scan be formed by associating colloidal particles with polymers. This may be nanometric silica particles, dispersed in water, and poly(ethylene oxide) [66–68]. Here, the shear-induced gelation and shear-induced flocculation are observed near the boundary of the phase separation region. Above a critical shear rate, the necklaces connect to each other to form threadlike objects which align along the velocity. At higher shear these objects associate sideways to form three-dimensional flocs.

With such systems, the viscosification under shear is reversible, but such associations are quite fragile and very sensitive to the ionic strength of the medium and to the presence of surfactants, which destroy the rheo-viscosifying properties. In addition, high molecular weight poly(oxyethylene) is very expensive and these polymers are manufactured in small quantities only, which is not favorable for industrial applications [69].

Copolymers of other type have been proposed. These consist of comonomers with little or no affinity for the silica and comonomers, which adsorb onto the silica. The latter comonomers carry functions which are capable of adsorbing onto the silica and thus contain one or more heteroatoms, with an electron pair. These monomers are listed in Table 70 and depicted in Figure 63.

Preferred copolymers are synthesized from acrylamide and vinylimidazole, or from acrylamide and vinylpyrrolidone. The copolymers are prepared by a radical polymerization.

Commercially available precipitated silica suspensions such as KLEBESOL® 30R25 and KLEBESOL® 30R9 can be used. On the other hand, systems prepared from Aerosil® silicas such as Aerosil® 380 hydrophilic silica do not exhibit a rheo-viscosifying phenomenon.

Table 70 Comonomers for rheo-viscosifying compositions [69]

No affinity to silica	Affinity to silica
Acrylamide	Vinylpyrrolidone
Acrylic acid	Vinylimidazole
Methacrylic acid	Vinylpyridine
2-Acrylamido-2-methylpropanesulfonate	Alkylvinylethers
Sulfonated styrene	N-Vinyl acetamide
	Hydroxyethyl methacrylate

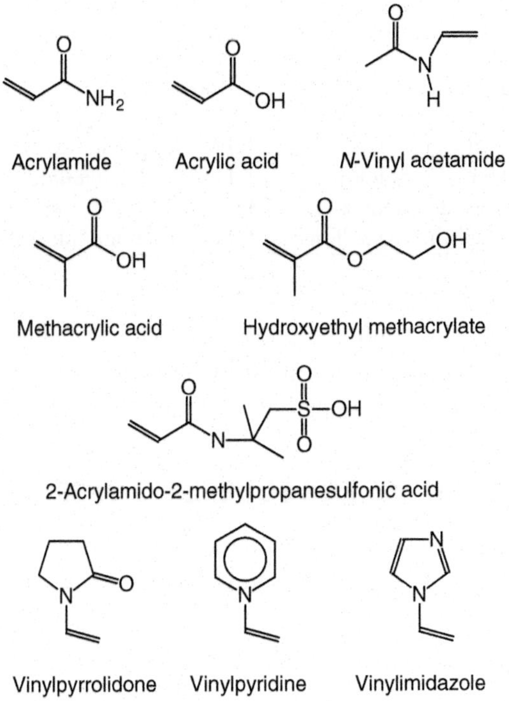

Acrylamide   Acrylic acid   N-Vinyl acetamide

Methacrylic acid   Hydroxyethyl methacrylate

2-Acrylamido-2-methylpropanesulfonic acid

Vinylpyrrolidone   Vinylpyridine   Vinylimidazole

Figure 63 Rheo-viscosifying monomers.

The viscosity increase or gelling of aqueous solutions with the copolymer/silica system is reversible, i.e., when the shear rate is reduced after a high shear rate applied, the viscosity reduces again. It has been observed that some gelling fluids exhibit a hysteresis effect, i.e., once gelling occurs, a much lower shear rate is sufficient to maintain the gelling effect for a particular period. Further reduction of the shear rate subsequently leads to a liquefaction of the fluid [69].

To these compositions, optionally a third additive a competing agent, can be added to the aqueous solution. The competing agent must adsorb onto the silica surface, either in a manner which is stronger than the adsorbable copolymer, or in a manner which is substantially equal or slightly weaker. In the first case, the goal is to reduce the available silica surface for adsorption of the copolymer by blocking a portion of the adsorption sites of the silica surface and by spreading out these sites. In contrast, in the second case, the competing agent reduces the intensity of adsorption of the copolymer by reversible competition for the silica

Table 71 Competing agents [69]

Compound	Compound
Alkylphenylethylene glycol	Butanol
Fatty alcohols	
Poly(ethylene oxide)	Poly(propylene oxide)
N-Methylpyrrolidone	Dimethylformamide
Isopropylacrylamide	
Polynaphthalene sulfonates	Polymelamine sulfonates

adsorption sites. Such a competing agent is selected from fatty alcohols nonionic surfactants, polyethers, N-substituted amides, and dispersing agents [69]. Specific examples of competing agents are given in Table 71.

10. SCALE INHIBITORS FOR SQUEEZING OPERATIONS

Squeezing is a three-stage process by which fluids are injected directly into the wellbore, reversing the flow of liquid back down into the reservoir [70].

First a dilute solution of scale inhibitor with surfactant is applied to clean and cool the near wellbore. This step is followed by a high concentration solution of the scale inhibitor—the pill, followed by a low concentration solution of inhibitor which is applied to move the pill away from the near wellbore, radially outward to a distance into the near wellbore which is designed to give maximum squeeze life.

The solutions are left in contact with the reservoir for 6-24 h to allow an adsorption equilibrium. Afterwards the well is returned to production. Adhesion to the formation allows the inhibitor to remain within the near-wellbore area without being pumped up in the oil/water emulsion, returning only at concentrations in the aqueous phase below 250 ppm for an extended period thus providing a longer treatment lifetime [70].

As a scale inhibitor for barium sulfate scale and calcium carbonate scale a water soluble polymer made from ethylenically unsaturated carboxylic acid is used. Examples of monomers are listed in Table 72 and shown in Figure 64.

Preferred monomers are acrylic acid, MA, or maleic acid anhydride. Phosphate-functional carboxylic acid polymers are effective at inhibiting calcium carbonate scale as well as barium sulfate scale. The phosphate moiety on the poly(acrylate) effects a combination of nucleation inhibition with crystal growth retardation and crystal growth modification. The addition of a sulfonic acid monomer to the polymer also allows excellent

Table 72 Ethylenically unsaturated carboxylic acids for scale inhibitors [70]

Acid monomer	Acid monomer
Acrylic acid	Angelic acid
Methacrylic acid	Cinnamic acid
Ethacrylic acid	p-Chloro cinnamic acid
α-Chloro-acrylic acid	Itaconic acid
α-Cyano acrylic acid	Citraconic acid
β-Methyl-acrylic acid	Mesaconic acid
α-Phenyl acrylic acid	Glutaconic acid
β-Acryloxy propionic acid	Aconitic acid
β-Styryl acrylic acid	Sorbic acid
Maleic acid	α-Chloro sorbic acid
Maleic acid anhydride	Fumaric acid
Vinylsulfonic acid	2-Acrylamido-2-methylpropane sulfonic acid
Styrene sulfonic acid	Ethylene glycol methacrylate phosphate

compatibility with the formation water, especially high calcium brines and effects a greater efficacy when encountering barium scales [70].

Desirable properties for a barium scale inhibitor are that the inhibitor should have a high salt tolerance, adsorb onto the oil-containing formation, not does desorb under high shear, is water soluble, and should be effective under high-temperature and high-pressure environments. An example for the preparation of a scale inhibitor is given below [70]

Preparation V–1: To a 2 l glass vessel equipped with stirrer, reflux condenser and means of temperature control, 200 g of propan-2-ol and 200 g of deionized water was charged then heated to reflux. A monomer mixture of acrylic acid (200 g), 2-acrylamido-2-methyl propane sulfonic acid (141.4 g), and ethylene glycol methacrylate phosphate (34.1 g) was fed over 3 h into the reactor. A initiator solution was fed concurrently with the monomer feed but with an overlap of 30 min and consisted of sodium persulfate (13.5 g), 35% hydrogen peroxide (55 g), and water (65 g). When both feeds were complete the reaction was held at reflux for 30 min then cooled. The propan-2-ol was removed by distillation on a rotary evaporator. The resulting polymer was neutralized with 50 g of 50% sodium hydroxide.■

11. MULTIFUNCTIONAL ADDITIVES

2,5-Dimercapto-1,3,4-thiadiazole compounds have long been used as metal passivators and load-carrying additives in lubricating oils and grease. The compounds exhibit excellent corrosion inhibiting and extreme-pressure properties.

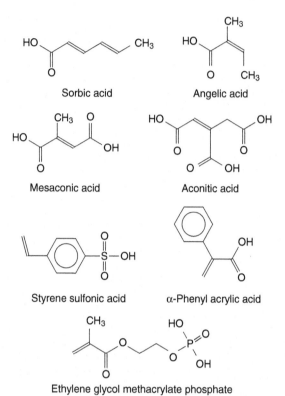

Figure 64 Monomers for scale inhibitors.

The tribological, anticorrosion, and antirust properties of 2,5-dimercapto-1,3,4-thiadiazole derivatives as water-soluble additives in a water-glycol hydraulic fluid have been investigated [71].

Their tribological properties were evaluated in detail by a four-ball wear test machine and an oscillating friction and wear tester. Also, their anticorrosion and antirust properties were investigated by copper strip corrosive tests and antirust tests, respectively. The worn surfaces were analyzed by scanning electron microscope and X-ray photonelectron spectroscope.

The 2,5-Dimercapto-1,3,4-thiadiazole compounds show an excellent solubility in the base liquid of a water-glycol hydraulic fluid. In addition, the antiwear and friction-reducing properties of the base liquid are improved, as well as copper corrosion inhibiting and antirust properties [71].

Tradenames appearing in the references are shown in Table 73.

Table 73 Tradenames in references

Tradename Description	Supplier
Aculyn™ (Series) hydrophobically-modified poly(acrylate) [57]	Rohm and Haas
Aquatreat® AR-545 Scale control agent [70]	Akzo Nobel
Imsil® A-15 Silica filler [29]	Unimin Corp.
MontBrite® 1240	Montgomery Chemicals
Aqueous solution of sodium borohydride and caustic soda [6]	
Polybor® Polymeric borate [57]	U.S. Borax of Valencia
Tergitol® (Series) Ethoxylated C11-15-secondary alcohols, surfactant [29]	Union Carbide Corp.
Texanol®	Eastman Chemical Comp.
2,2,4-Trimethyl-1,3-pentanediol monoisobutyrate [29]	
Ti-Pure® Titanium dioxide [29]	DuPont
Triton® X (Series)	Union Carbide Corp. (Rohm & Haas)
Poly(alkylene oxide), nonionic surfactants [29]	
Triton® X-100 Hydrophilic poly(ethylene oxide) [29]	Dow Chemical Comp.

REFERENCES

[1] Chatterji J, Borchardt JK. Applications of water-soluble polymers in the oil field. J Petrol Technol 1981;33(11). doi:10.2118/9288-PA.

[2] Stahl GA, Schulz DN. Water-soluble polymers for petroleum recovery. New York: Springer; 1988.

[3] Taylor KC, Nasr-El-Din HA. Water-soluble hydrophobically associating polymers for improved oil recovery: a literature review. Society of Petroleum Engineers. ISBN 9781555634575; 1995. doi:10.2118/29008-MS.

[4] Taylor KC, Nasr-El-Din HA. Water-soluble hydrophobically associating polymers for improved oil recovery: a literature review. J Petrol Sci Eng 1998;19(3-4):265-80. URL: http://www.sciencedirect.com/science/article/pii/S092041059700048X. doi:10.1016/S0920-4105(97)00048-X.

[5] Totten GE, DeNegri VJ, editors. Handbook of hydraulic fluid technology. 2nd ed. Boca Raton, FL: Taylor & Francis; 2012. ISBN 9781420085266.

[6]   Lumsden CA, Diaz RO. Method and composition for oil enhanced recovery. US Patent 8662171, assigned to Montgomery Chemicals, LLC (Conshohocken, PA) Nalco Company (Naperville, IL); 2014. URL: http://www.freepatentsonline.com/8662171.html.

[7]   Tiwari L, Meyer GR, Horsup D. Environmentally friendly bis-quaternary compounds for inhibiting corrosion and removing hydrocarbonaceous deposits in oil and gas applications. US Patent 7951754, assigned to Nalco Company (Naperville, IL); 2011. URL: http://www.freepatentsonline.com/7951754.html.

[8]   Tiwari L. Mono and bis-ester derivatives of pyridinium and quinolinium compounds as environmentally friendly corrosion inhibitors. US Patent 8585930, assigned to Nalco Company (Naperville, IL); 2013. URL: http://www.freepatentsonline.com/8585930.html.

[9]   Zaid GH, Wolf BA. Well treatment composition for use in iron-rich environments. US Patent 6720291, assigned to Jacam Chemicals, L.L.C. (Sterling, KS); 2004. URL: http://www.freepatentsonline.com/6720291.html.

[10]   Sitz C, Frenier W, Vallejo C. Acid corrosion inhibitors with improved environmental profiles. In: Proceedings of SPE international conference and exhibition on oilfield corrosion; 2012. doi:10.2118/155966-ms.

[11]   Mannich   C,   Krösche   W.   Über   ein   Kondensationsprodukt   aus Formaldehyd, Ammoniak und Antipyrin. Arch Pharm 1912;250(1):647-67. doi:10.1002/ardp.19122500151.

[12]   Martin JA, Valone FW. The existence of imidazoline corrosion inhibitors. Corrosion 1985;41(5):281-7. doi:10.5006/1.3582003.

[13]   Edwards A, Osborne C, Webster S, Klenerman D, Joseph M, Ostovar P, et al. Mechanistic studies of the corrosion inhibitor oleic imidazoline. Corros Sci 1994;36(2):315-25. doi:10.1016/0010-938x(94)90160-0.

[14]   Meyer GR. Imidazoline corrosion inhibitors. US Patent 7057050, assigned to Nalco Energy Services L.P. (Sugar Land, TX); 2006. URL: http://www.freepatentsonline.com/7057050.html.

[15]   Naragli A, Obeyesekere NU. Corrosion inhibitors with low environmental toxicity. US Patent 6475431, assigned to Champion Technologies, Inc. (Houston, TX); 2002.

[16]   Clewlow PJ, Haselgrave JA, Carruthers N, O'Brien TM. Corrosion inhibitors. US Patent 5300235, assigned to Exxon Chemical Patents Inc. (Linden, NJ); 1994. URL: http://www.freepatentsonline.com/5300235.html.

[17]   Pou TE, Fouquay S. Polymethylenepolyamine dipropionamides as environmentally safe inhibitors of the carbon corrosion of iron. US Patent 6365100, assigned to CECA, S.A. (Puteaux, FR); 2002. URL: http://www.freepatentsonline.com/6365100.html.

[18]   Talbot RE, Jones CR, Hills E. Scale inhibition in water systems. US Patent 7572381, assigned to Rhodia U.K. Limited (Hertfordshire, GB); 2009. URL: http://www.freepatentsonline.com/7572381.html.

[19]   Viloria A, Castillo L, Garcia JA, Biomorgi J. Aloe derived scale inhibitor. US Patent 7645722, assigned to Intevep, S.A. (Caracas, VE); 2010. URL: http://www.freepatentsonline.com/7645722.html.

[20]   Stefl BA, George KL. Antifreezes and deicing fluids. In: Kirk-Othmer Encyclopedia of Chemical Technology, vol. 3. New York: John Wiley and Sons; 1996, p. 347-66.

[21]   Barannik VP, Kubyshkina EK, Lezina NM. Corrosion and thermophysical properties of lithium chloride-based coolant. Zashch Korrozii Okhrana Okruzhayushchej Sredy 1995;(8-9):12-4.

[22]   Darden JW, McEntire EE. Dicyclopentadiene dicarboxylic acid salts as corrosion inhibitors. EP Patent 0200850 assigned to Texaco Development Corporation; 1986. URL: https://www.google.at/patents/EP0200850A1?cl=en.

[23] Darden JW, Triebel CA, Van Neste WA, Maes JP. Monobasic-dibasic acid/salt antifreeze corrosion inhibitor. EP Patent 0229440 assigned to Texaco Development Corporation, S.A. Texaco Belgium N.V.; 1990. URL: https://www.google.at/patents/EP0229440B1?cl=en.

[24] Hohlfeld R. Longer life for glyco-based stationary engine coolants. Pipeline Gas J 1996;223(7):55-7.

[25] Sapienza R, Johnson A, Ricks W. Environmentally benign anti-icing or deicing fluids employing triglyceride processing by-products. US Patent 7270768, assigned to MLI Associates, LLC (Westerville, OH); 2007. URL: http://www.freepatentsonline.com/7270768.html.

[26] Luo J, Zhu H, Bai F, Wang P, Ding B, Yang J, et al. Dendritic comb-shaped polymer thickening agent, preparation of the same and application thereof. US Patent Application 20130090269, assigned to Petrochina Company Limited, Beijing (CN); 2013. URL: http://www.freepatentsonline.com/20130090269.html.

[27] Subramanian S, Burgazli C, Zhu YP, Zhu S, Feuerbacher D. Quaternary ammonium salts as thickening agents for aqueous systems. US Patent 7776798, assigned to Akzo Nobel Surface Chemistry LLC (Chicago, IL); 2010. URL: http://www.freepatentsonline.com/7776798.html.

[28] Müller H, Herold CP, van Tapavizca S, Dolhaine H, von Rybinski W, Wichelhaus W. Water based drilling and well-servicing fluids with swellable, synthetic layer silicates. US Patent 4888120, assigned to Henkel Kommanditgesellschaft auf Aktien (Düsseldorf, DE); 1989. URL: http://www.freepatentsonline.com/4888120.html.

[29] Chaiko DJ. Process for preparing organoclays for aqueous and polar-organic systems. US Patent 6172121, assigned to The University of Chicago (Chicago, IL); 2001. URL: http://www.freepatentsonline.com/6172121.html.

[30] Müller H, Maker D. Emulsion-based cleaning composition for oilfield applications. US Patent 8763724, assigned to Emery Oleochemicals GmbH (DE); 2014. URL: http://www.freepatentsonline.com/8763724.html.

[31] Nunes DG, da Silva AdP, Cajaiba J, Pérez-Gramatges A, Lachter ER, Nascimento RSV. Influence of glycerides-xanthan gum synergy on their performance as lubricants for water-based drilling fluids. J Appl Polym Sci 2014. doi:10.1002/app.41085.

[32] Abdo J. Nano-attapulgite for improved tribological properties of drilling fluids. Surf Interface Anal 2014. doi:10.1002/sia.5472.

[33] Wu Y. Corrosion inhibitor for wellbore applications. US Patent 5654260, assigned to Phillips Petroleum Company (Bartlesville, OK); 1997. URL: http://www.freepatentsonline.com/5654260.html.

[34] Pakulski M, Hlidek BT. Slurried polymer foam system and method for the use thereof. US Patent 5360558, assigned to The Western Company of North America (Houston, TX); 1994. URL: http://www.freepatentsonline.com/5360558.html.

[35] Chatterji J, King BJ, Cronwell RS, Brenneis DC, Gray DW. Foamed cement slurries, additives and methods. US Patent 6951249, assigned to Halliburton Energy Services, Inc. (Duncan, OK); 2005. URL: http://www.freepatentsonline.com/6951249.html.

[36] Szymanski MJ, Lewis SJ, Wilson JM, Karcher AL. Cement compositions comprising environmentally compatible defoamers and methods of use. US Patent 7150322, assigned to Halliburton Energy Services, Inc. (Duncan, OK); 2006.

[37] Gioia F, Urciuolo M. The containment of oil spills in unconsolidated granular porous media using xanthan/cr(iii) and xanthan/al(iii) gels. J Hazard Mater 2004;116(1-2):83-93.

[38] Ostryanskaya GM, Abramov YD, Makarov VN, Osipov SN, Razhkevich AV. Gel-forming plugging composition—contains ligno-sulphonate, modified carboxymethyl cellulose, bichromate, calcium chloride and water. SU Patent 1776766; 1992.

[39] Dobroskok BE, Gulyaeva ZG, Kubareva NN, Muslimov RK, Nizova SA, Terekhova VV, et al. Plugging composition for hydro-insulation of oil stratum—contains polydimethyl-diallyl ammonium chloride, sodium salt of carboxy methyl cellulose, sodium chloride and water. SU Patent 1758209, assigned to Moscow Gubkin Oil Gas Inst.; 1992.

[40] Merrill LS. Fiber reinforced gel for use in subterranean treatment process. WO Patent 9319282, assigned to Marathon Oil Co.; 1993.

[41] Perejma AA, Pertseva LV. Complex reagent for treating plugging solutions—comprises hydrolysed polyacrylonitrile, ferrochromolignosulphonate cr-containing additive, waste from lanolin production treated with triethanolamine and water. RU Patent 2013524, assigned to N Caucasus Nat Gaz Res.; 1994.

[42] Moradi-Araghi A. Gelling compositions useful for oil field applications. US Patent 5432153, assigned to Phillips Petroleum Company (Bartlesville, OK); 1995. URL: http://www.freepatentsonline.com/5432153.html.

[43] Lyadov BS. Gel-forming composition for isolating works in well—contains polyacrylamide, urotropin, water-soluble chromate(s) and water. SU Patent 1730432, assigned to Borehole Consolidation Mu; 1992.

[44] Merrill LS. Fiber reinforced gel for use in subterranean treatment process. GB Patent 2277112; 1994.

[45] Merrill LS. Fiber reinforced gel for use in subterranean treatment processes. US Patent 5377760, assigned to Marathon Oil Company (Findlay, OH); 1995. URL: http://www.freepatentsonline.com/5377760.html.

[46] Kosyak SV, Danyushevskij VS, Pshebishevskij ME, Trapeznikov AA. Plugging formation fluid transmitting channel—by successive injection of aqueous solution of polyacrylamide and liquid glass, buffer liquid and aqueous solution of polyacrylamide and manganese nitrate. SU Patent 1797645; 1993.

[47] Kotelnikov VS, Demochko SN, Fil VG, Rybchich II. Polymeric composition for isolation of absorbing strata—contains ferrochromo-lignosulphonate, water-soluble acrylic polymer and water. SU Patent 1730435, assigned to Ukr. Natural Gas Res. Inst.; 1992.

[48] Mumallah NAK, Shioyama TK. Process for preparing a stabilized chromium (iii) propionate solution and formation treatment with a so prepared solution. EP Patent 0194596 assigned to Phillips Petroleum Company; 1986. URL: https://www.google.at/patents/EP0194596A2?cl=en.

[49] Mumallah NA. Chromium (iii) propionate: a crosslinking agent for water-soluble polymers in real oilfield waters. In: Proceedings volume. SPE Oilfield Chem. Int. Symp. (San Antonio, 2/4-6/87); 1987.

[50] Nanda SK, Kumar R, Goyal KL, Sindhwani KL. Characterization of polyacrylamine-Cr^{6+} gels used for reducing water/oil ratio. In: Proceedings volume. SPE Oilfield Chem. Int. Symp. (San Antonio, 2/4-6/87); 1987.

[51] Hanes Jr RE, Weaver JD, Slabaugh BF, Barrick DM. Environmentally benign viscous well treating fluids and methods. US Patent 7000702, assigned to Halliburton Energy Services, Inc. (Duncan, OK); 2006. URL: http://www.freepatentsonline.com/7000702.html.

[52] Todd BL, Frost KA. Methods and compositions for treating subterranean zones using environmentally safe polymer breakers. US Patent 6918445, assigned to Halliburton Energy Services, Inc. (Duncan, OK); 2005. URL: http://www.freepatentsonline.com/6918445.html.

[53] Smith JE. Performance of 18 polymers in aluminum citrate colloidal dispersion gels. In: Proceedings volume. SPE Oilfield Chem. Int. Symp. (San Antonio, 2/14-17/95); 1995, p. 461-70.

[54]   Rocha CA, Green DW, Willhite GP, Michnick MJ. An experimental study of the interactions of aluminum citrate solutions and silica sand. In: Proceedings volume. SPE Oilfield Chem. Int. Symp. (Houston, 2/8-10/89); 1989, p. 403-13.

[55]   Avakov VE. Preventing absorption of drilling solution—by introduction of water-expandable material based on bentonite clay and polyacrylamide into circulating drilling solution. SU Patent 1745123; 1992.

[56]   Javora P, Wang X, Stevens R, Pearcy R. Water based insulating fluids for deepwater riser applications. SPE Projects Facilities & Construction 2006;1(1). doi:10.2118/88547-PA.

[57]   Wang X, Qu Q, Dawson JC, Gupta DVS. Thermal insulation compositions containing organic solvent and gelling agent and methods of using the same. US Patent 7713917, assigned to BJ Services Company (Houston, TX); 2010. URL: http://www.freepatentsonline.com/7713917.html.

[58]   Dzialowski A, Ullmann H, Sele A, Oosterkamp LD. The development and application of environmentally acceptable thermal insulation fluids. In: SPE/IADC Drilling Conference. SPE-79841-MS; SPE/IADC Drilling Conference, 19-21 February, Amsterdam, Netherlands. The Woodlands, TX: Society of Petroleum Engineers; 2003. doi:10.2118/79841-MS.

[59]   Anonymous. Engineering and operating guide for DOWFROST and DOWFROST HD inhibited propylene glycol-based heat transfer fluids. Technical Guide 180-01286-0208 AMS; Dow Chemical Company; Midland, MI; 2008. Table 18, p. 27. URL: http://msdssearch.dow.com/PublishedLiteratureDOWCOM/dh_010e/0901b8038010e417.pdf?filepath=heattrans/pdfs/noreg/180-01286.pdf&fromPage=GetDoc.

[60]   Parusyuk AV, Galantsev IN, Sukhanov VN, Ismagilov TA, Telin AG, Barinova LN, et al. Gel-forming compositions for leveling of injectivity profile and selective water inflow shutoff. Neft Khoz 1994;(2):64-8.

[61]   Kuznetsov VL, Lyubitskaya GA, Kolesnik EI, Kazakova EN, Kurochkin BM, Lobanova VN. Plugging solution for isolating absorption zones in oil and gas wells—contains prescribed synthetic latex, water soluble salt of methacrylate-methacrylic acid copolymer as additive, and water. RU Patent 2024734, assigned to Drilling Tech. Res. Inst.; 1994.

[62]   Soreau M, Siegel D. Injection composition for filling or reinforcing grounds. GB Patent 2170838; 1986.

[63]   Soreau M, Siegel D. Application of a gelling agent for an alkali-silicate solution for sealing and consolidating soils (Verwendung eines Geliermittels für zum Abdichten und Verfestigen von Böden bestimmte Alkalisilikatlösung). DE Patent 3506095, assigned to Soc. Francaise Hoechst; 1990.

[64]   Sharif S. Borate cross-linking solutions. US Patent 5310489, assigned to Zirconium Technology Corporation (Midland, TX); 1994. URL: http://www.freepatentsonline.com/5310489.html.

[65]   Mondshine TC. Crosslinked fracturing fluids. US Patent 4619776, assigned to Texas United Chemical Corp. (Houston, TX); 1986. URL: http://www.freepatentsonline.com/4619776.html.

[66]   Liu SF, Lafuma F, Audebert R. Rheological behavior of moderately concentrated silica suspensions in the presence of adsorbed poly(ethylene oxide). Colloid Polym Sci 1994;272(2):196-203. doi:10.1007/BF00658848.

[67]   Liu S, Legrand V, Gourmand M, Lafuma F, Audebert R. General phase and rheological behavior of silica/peo/water systems. Colloids Surf A Physicochem Eng Asp 1996;111(1-2):139-45. doi:10.1016/0927-7757(95)03500-1.

[68]   Cabane B, Wong K, Lindner P, Lafuma F. Shear induced gelation of colloidal dispersions. J Rheol 1997;41(3):531. doi:10.1122/1.550874.

[69] Maroy P, Lafuma F, Simonet C. Liquid compositions which reversibly viscosify or gel under the effect of shear. US Patent 6586371, assigned to Schlumberger Technology Corporation (Sugar Land, TX); 2003. URL: http://www.freepatentsonline.com/6586371.html.

[70] Crossman M, Holt SPR. Scale control composition for high scaling environments. US Patent 6995120, assigned to National Starch and Chemical Investment Holding Corporation (New Castle, DE); 2006. URL: http://www.freepatentsonline.com/6995120.html.

[71] Wang J, Wang J, Li C, Zhao G, Wang X. A study of 2,5-dimercapto-1,3,4-thiadiazole derivatives as multifunctional additives in water-based hydraulic fluid. Ind Lubr Tribol 2014;66(3):402-10. doi:10.1108/ilt-11-2011-0094.

CHAPTER VI

Environmental Aspects and Waste Disposal

1. ENVIRONMENTAL REGULATIONS

In response to effluent limitation guidelines promulgated by the Environmental Protection Agency for discharge of drilling wastes offshore, alternatives to water-based drilling muds (WBMs) and oil-based drilling muds (OBMs) have been developed. Thus synthetic-based muds are more efficient than WBMs for drilling difficult and complex formation intervals and have lower toxicity and smaller environmental impacts than diesel or conventional mineral OBMs.

Synthetic drilling fluids may present a significant pollution prevention opportunity because the fluids are recycled, and smaller volumes of metals are discharged with the cuttings than for WBMs. A framework for a comparative risk assessment for the discharge of synthetic drilling fluids has been developed. The framework will help identify potential impacts and benefits associated with the use of drilling muds [1].

1.1 The Clean Water Act

In 1972, in the midst of a national concern about untreated sewage, industrial and toxic discharges, destruction of wetlands, and contaminated runoff, the principal law to protect the nation's waters was passed [2].

Originally enacted in 1948 to control water pollution primarily based on state and local efforts, the Federal Water Pollution Control Act, or Clean Water Act (CWA), was totally revised in 1972 to give the Act its current shape. The CWA set a new national goal to restore and maintain the chemical, physical, and biological integrity of the Nation's waters, with interim goals that all waters be fishable and swimmable where possible. The Act embodied a new federal-state partnership, where federal guidelines, objectives and limits were to be set under the authority of the US Environmental Protection Agency, while states, territories, and authorized tribes would largely administer and enforce the CWA programs, with significant federal technical and financial assistance. The Act also gave citizens a strong role to play in protecting and restoring waters.

Water-Based Chemicals and Technology for Drilling,
Completion, and Workover Fluids
http://dx.doi.org/10.1016/B978-0-12-802505-5.00006-8

Copyright © 2015 Elsevier Inc.
All rights reserved.

The CWA specifies that all discharges into the nation's waters are unlawful unless authorized by a permit and sets baseline, across-the-board technology-based controls for municipalities and industry. It requires all dischargers to meet additional, stricter pollutant controls where needed to meet water quality targets and requires federal approval of these standards. It also protects wetlands by requiring *dredge and fill* permits. The CWA authorizes federal financial assistance to states and municipalities to help achieve these national water goals. The Act has robust enforcement provisions and gives citizens a strong role to play in watershed protection. Congress has revised the Act, most notably in 1987, where it established a comprehensive program for controlling toxic pollutants and stormwater discharges, directed states to develop and implement voluntary nonpoint pollution management programs, and encouraged states to pursue groundwater protection. Notwithstanding these improvements, the 1972 statute, its regulatory provisions and the institutions that were created 40 years ago, still make up the bulk of the framework for protecting and restoring the nation's rivers, streams, lakes, wetlands, and coastal waters [2].

1.2 Safe Drinking Water Act

The Safe Drinking Water Act (SDWA) is the main federal law that ensures the quality of Americans' drinking water [3]. Under SDWA, EPA sets standards for drinking water quality and oversees the states, localities, and water suppliers who implement those standards.

The SDWA was originally passed by Congress in 1974 to protect public health by regulating the nation's public drinking water supply. The law was amended in 1986 and 1996 and requires many actions to protect drinking water and its sources: rivers, lakes, reservoirs, springs, and ground water wells.

1.3 Regulatory Framework of Fracturing

The regulatory framework of fracturing has been reviewed [4]. The current Federal regulations on fracturing are derived from the SDWA [3] and the CWA [2].

Federal, State, and local regulations typically apply to hydraulic fracturing, but there are various loopholes purposefully created by governmental actors to enable the drilling and gas industry access to resources with minimal interference.

Currently, the Federal regulations only contain protections that existed prior to the advent of horizontal well drilling and local permitting seems to be perfunctory at best.

This lack of regulation or oversight may well be due to the inability of government to keep up with the pace of innovation in the energy industry or it could be seen as a play for jobs and economic growth. An example of this can be seen in Pennsylvania, where Marcellus Shale exploitation is in full swing and has been for several years [4].

2. TOXICITY TESTS

2.1 Sediment Toxicity Test

A sediment-toxicity test was required by the US Environmental Protection Agency in addition to the existing water–column test to define a best-available-technology limit [5].

To fulfill these demands, an interindustry research group worked with the US Environmental Protection Agency to develop a suitable test that can meet a technology-based discharge standard. The toxicity of a discharged field drilling fluid is compared to a reference synthetic-based drilling fluid with $C_{16}-C_{18}$ internal olefin. The results of this program have been reviewed [5].

2.2 Mysid Acute Toxicity Test

Standard acute toxicity tests were conducted on juvenile shrimp and small fish that are found in the gulf and are commonly used in toxicity testing. The tests were conducted on mixtures of Louisiana Sweet Crude Oil and eight different dispersant products found on the National Contingency Plan Product Schedule—Dispersit SPC 1000, Nokomis 3-F4, Nokomis 3-AA, ZI-400, SAFRON Gold, Sea Brat #4, Corexit 9500 A, and JD 2000. The same eight dispersants were used during EPA's first round of independent toxicity testing.

All eight dispersants were found to be less toxic than the dispersant-oil mixture to both test species. Louisiana Sweet Crude Oil was more toxic to mysid shrimp than the eight dispersants when tested alone. Oil alone had similar toxicity to mysid shrimp as the dispersant-oil mixtures, with exception of the mixture of Nokomis 3-AA and oil, which was found to be more toxic than oil [6].

2.3 Environmental Tests

When determining toxicity of materials used in conjunction with offshore drilling and production activities, Canada requires that the proposed materials undergo the Microtox® acute toxicity test [7]. The test has been described elsewhere [8, 9].

The Microtox® acute toxicity test operates on the basis of monitoring the level of light emission from luminescent bacteria. Luminescent bacteria produce light as a byproduct of their cellular respiration. Exposure to toxic conditions result in a decrease in the rate of respiration, thereby reducing the rate of luminescence. Consequently, toxicity is measured as a percentage of luminescence lost. The test endpoint is measured as the effective concentration (EC) of a test sample that reduces light emission by a specific amount under defined conditions of time and temperature. Generally, the effective concentration is expressed as $EC_{50}(15)$, which is the effective concentration of a sample which reduces light emission by 50% at 15 min at 15 °C. The length of time of exposure, and the minimum EC_{50} values will vary depending on the local legislation. In some embodiments, the additives of the present invention are added to a water-based wellbore fluid in concentrations resulting in $EC_{50}(15)$ values greater than 50%. In other embodiments, $EC_{50}(15)$ values greater than 70% or $EC_{50}(15)$ values greater than 90% are demanded [7].

3. POLLUTANTS

3.1 Toxicological Data

Toxicological data for some substances are collected in Table 74.

3.2 Mercaptans

Mercaptans are used for various purposes, such as biocorrosion inhibitors, anti-fungal drugs in medical applications coating agents for metallic surfaces, and as vulcanization accelerators for rubbers [10, 11].

The photocatalytic oxidation of aqueous solutions of 2-mercaptobenzo-thiazole, 2-mercaptobenzimidazole, 2-mercaptobenzoxazole, 2-mercapto-pyridine, and p-toluylmercaptan was tested using a heterogeneous catalyst of TiO_2 [12].

In order to optimize the degradation process, the amount of catalyst, flow of oxygen, pH, and time of irradiation were varied. The above mentioned mercaptan derivatives are almost completely mineralized to carbon dioxide, ammonia, and sulfate ion by this photocatalytic method [12].

Table 74 Toxicological data

Compound	CAS	Acute oral LD50/[mg kg^{-1}]	Animal
Acrylic acid	79-10-7	830	Mouse
2-Acrylamido-2-methylpropane sulfonic acid	15214-89-8	1440	Rat
Ammonium persulfate	7727-54-0	689	Rat
Choline chloride	67-48-1	3400	Rat
2,2'-Azobisisobutyronitrile	78-67-1	100	Rat
Diethanol amine	111-42-2	710	Rat
2-Aminoethanol	141-43-5	1720	Rat
Benzoic acid	65-85-0	1700	Rat
Dimethylurea	96-31-1	4000	Rat
Citric acid	77-92-9	3000	Rat
Ethylene glycol	107-21-1	4700	Rat
Ethylene diamine tetraacetic acid	60-00-4	4500	Rat
Ethylenediaminetetraacetate dipotassium salt	25102-12-9	2150	Rat
Formic acid	64-18-6	1100	Rat
Glutaraldehyde 50% in water		134	Rat
Glyoxal 40% in water		1100	Rat
2-Mercaptobenzothiazole	149-30-4	100	Rat
Methanol	67-56-1	5628	Rat
Sodium perborate tetrahydrate	10486-00-7	1200	Rat
Sulfamic acid	5329-14-6	3160	Rat
Triazol	288-88-0	1750	Rat

3.2.1 Mercaptobenzothiazole

2-Mercaptobenzothiazole is known as a widespread used, but toxic and poorly biodegradable pollutant [13].

3.3 Allylamine

Certain aspects of the toxicity of allylamines have been known for a long time [14]. Allylamine is toxic to the cardiovascular system causing aortic, valvular, and myocardial lesions. The acute toxicity is believed to emerge from the metabolism of allylamine to acrolein, which is highly reactive [15].

4. USE OF WASTE BYPRODUCTS AS ADDITIVES

Many of the organic drilling fluid additives are waste byproducts [16]. Thus, their use as drilling fluid additives has the advantage of providing a use for a waste product. The use of biodegradable organic additives makes the disposal of waste drilling fluids less expensive and less of an environmental concern.

Olive oil is widely used as a food additive and is produced by pressing olives. The pressing of olives to remove olive oil produces large quantities of waste olive pulp which must be disposed properly.

The disposal has been accomplished by burning the waste olive pulp. However, tighter environmental laws have decreased the feasibility of burning waste olive pulp for disposal purposes. Accordingly, an environmentally safe alternative for the disposal of olive pulp is desirable.

The use of olive pulp as an additive to drilling fluid provides an environmentally safe method of disposal of significant quantities of olive pulp, while also providing drilling fluids which prevent fluid loss, lost circulation, adverse effects on formations, and undesirable chemical reactions in the wellbore [16].

Ground olive pulp is prepared by drying olive pulp to remove the residual water, then grinding the pulp to a particle size of less than about 1500 μm. After drying, the olive pulp contains some 14% of residual olive oil.

The ground olive pulp can be added to oil- or water-based drilling fluids. In this way, the preparation and use of waste olive pulp as a drilling fluid additive provides an environmentally safe method for disposing of waste olive pulp.

During drilling operations or completion procedures, the olive pulp particles enter the pores of sands and micro fractures in the ground formation, plugging off leaks. The particles may then swell, thus further sealing off the formation [16]. Also, the residual olive oil in the olive pulp is believed to contribute to the lubrication properties of the drilling fluid.

5. WASTE STREAMS

A study has examined the waste streams from horizontal hydraulic fracture wells in the Marcellus formation [17]. Protective measures have been proposed that would minimize exposure.

The results showed that flowback, drilling muds, and horizontal hydraulic fracture fluids all exceeded the SDWA [3] limits to varying degrees.

Due to the contaminants found in these substances, proper handling and containment is essential to prevent harm to the environment [17].

6. DISCHARGE OF WASTE FLUIDS IN CEMENT

Waste fluids are commonly produced or recovered during drilling, transportation, or storage operations [18]. These fluids may arise, in the course of well stimulation, acid flowback, initial well flowback, completions, acid mine drainage, or pipeline maintenance. In general, there is a need to dispose the produced water. If the produced water has to be managed at the surface for processing and treatment the needed procedures are often expensive, time-consuming, and complicated. Therefore, improved methods for the disposal of produced water are highly desirable.

The input of drilling fluid wastes into a cement composition for the purpose of disposal has been known and assessed in the early prior art [19].

However, early efforts to push a drilling fluid into cement materials have posed certain problems such as increased viscosity and flocculation when the cementitious material is added and pumped down.

A cost effective and reliable technique for the immobilization of large amounts of drill cuttings is the inclusion into lime, pozzolanas, Portland, or slag cement. Unfortunately, chloride ions affect the setting of the cement as these ions act as retarders for setting. Moreover, the mechanical strength is reduced. For this reason, the disposal sodium chloride containing drill cutting wastes in the way sketched above is still problematic.

However, the addition of orthophosphate seems to form a continuous and weakly soluble network in the cement matrix, which reduces the release of the salt. Actually, apatite and hydrocalumite are formed. These phases encapsulate the salt grains within a network, thus lowering its interaction with water or trap the chloride [20]. Chloride trapping into hydrocalumite in ordinary Portland cement has been reported [21]. At high pH, hydrocalumite precipitates according to

$$2Ca(OH)_2 + Al(OH)_4^- + Cl^- + 2H^+ \rightarrow Ca_2(OH)_6Cl \cdot 3H_2O. \quad (11)$$

By leaching experiments at cement matrixes, where oil-based cuttings were embedded, it has been shown that the treatment of the cuttings with potassium phosphate strongly decreases the amount of dissolved salt from 41.3% to 19.1%. In contrast, aluminum phosphate is more efficient for the stabilization of water-based cuttings [20].

Methods of treating synthetic drill cuttings intended to landfills or for potential reuse as construction products have been screened [22]. Two synthetic mixes were used based on average concentrations of specific contaminates present in typical drill cuttings from the North Sea and the Red Sea areas. The two synthetic drill cuttings contained similar chloride content of 2.03% and 2.13% but different hydrocarbon content of 4.20% and 10.95%, respectively. Hence, the mixes were denoted as low and high oil content mixes, respectively.

A number of conventional binders for stabilization and solidification have been tested, including Portland cement, lime, and blast furnace slag. In addition, some novel binders such as microsilica and magnesium oxide cement have been screened. Despite of differences in the hydrocarbon content in the synthetic cuttings under investigation, the measured mechanical properties of the samples with the same binder type and content were similar.

Moreover, experiments of the leachability of the samples showed a reduction of the amount leached into a stable nonreactive hazardous waste. Leaching tests are standardized by an European standard [23]; however, there are still other similar tests that do the job [22].

Experiments of the leachability of paraffins revealed that among several binders, lime-Portland cement binders showed the best performance. Even 10% of binder reduced the leachability in comparison to untreated drill cuttings by more than 85% [22].

Converted drilling fluid compositions may exhibit some gellation and are particularly temperature sensitive [24]. In other words, if wellbore temperatures exceed a predetermined level, the cement composition has a tendency to set or harden rapidly. Since wellbore temperature conditions are difficult to control or predict in many instances, a reduced temperature sensitivity of the drilling fluid converted to cement is highly desirable [19].

In order to dispose the produced water recovered during hydrocarbon drilling, production, the method produced water is collected and in time it is mixed with Type I Portland cement to form a cementitious slurry. Potential additives to those compositions are set-accelerators, fluid loss control additives, anti-foam agents, waterproofing agents, dispersing agents, and plasticizing agents.

This slurry is then introduced into an underground void and allowed to set and form a cement. It has been demonstrated that an acceptable compressive strength and minimal resulting free water can be attained by mixing produced water with Type I Portland Cement [18].

6.1 Conversion into Cements

Water-based drilling fluids may be converted into cements using a hydraulic blast furnace slag [25–29]. Hydraulic blast furnace slag is a unique material that has low impact on rheologic and fluid loss properties of drilling fluids. It can be activated to set in drilling fluids that are difficult to convert to cements with other solidification technologies.

Hydraulic blast furnace slag is a more uniform and consistent quality product than are Portland well cements, and it is available in large quantities from multiple sources. Fluid and hardened solid properties of blast furnace slag and drilling fluids mixtures used for cementing operations are comparable with the properties of conventional Portland cement compositions.

In the manufacture of Portland cement, many otherwise-waste materials can be used either as a substitute for the traditional raw material, or as a secondary fuel (e.g., used tires) [30, 31]. In particular, drilling wastes can be introduced in the clinker burning process [32].

For both waste disposal and cement manufacturers, a mutual benefit will emerge. The cement manufacturing companies reduce their demand for traditional raw materials and save the limited capacity of landfills and other waste-treatment industries.

Waste water-based drilling fluids can be solidified by adding cement mixtures [33], in particular, those with low-quality blast furnace slags [26, 28, 34, 35]. Such mixtures have already been applied in wells at temperatures from approximately 4 to 315 °C. The disposal of rock cuttings is achieved by [36]:

- Combining the cuttings with water and blast furnace slag,
- Injecting the slurry into the annulus surrounding a wellbore casing, and
- Solidifying the cuttings, water, and slag.

Solidification in blast furnace slag cement is inexpensive [36]. The slag is compatible with both oil-based and WBM. Drilling fluids, therefore, do not need to be removed from the drilling cuttings prior to solidification in the wellbore annulus [37].

7. DISCHARGE IN POLYMERS

A method of fabricating a composite material from polymers and borehole solids has been described [38]. The borehole solids typically contain oil, particularly on the surface.

The borehole liquid being present on the borehole solids includes or consists of water and also residual oil. Thus, the oil containing borehole solids may be preprocessed before becoming embedded or also may be directly embedded in the base material without preprocessing.

Preprocessing the borehole solids includes chemically treating the borehole solids to thereby alter at least part of the borehole liquid. A physical treatment of the borehole solids may consist of a heat treatment. Also, preprocessing of the borehole solids is done by suspending the borehole solids in a mixture of solvents. A mixture of solvents may include a polar solvent and a non polar solvent, e.g., water and ethanol.

If the types of minerals of the borehole solids are very similar to the traditionally used fillers for polymers, similar achievements can be reached by using borehole solids as polymer fillers. Since the properties of the polymer filler composition are depending mostly on the polymer type and on the amount of filler and its particle size, shape, or the functional surface, almost all type of borehole solids can be used as fillers.

The polymeric base material includes at least one of a semicrystalline polymer and an amorphous polymer. The semicrystalline polymer may be poly(propylene). The amorphous polymer may be poly(styrene) (PS). The mixing of polymer and drilling waste can be performed with a twin screw compounder or a co-kneader. The above detailed method allows the recycling of oil-contaminated borehole solids while providing a resource for a filler for polymers [38].

8. WATER-BASED DRILLING FLUID DISPOSAL

8.1 Encapsulation

A drilling mud can be disposed of in an environmentally sound manner by mixing the mud with a crosslinkable polymer and a crosslinking agent to form a composition that solidifies at a predetermined time [39].

Before curing, the composition is injected into a subterranean formation, preferably into an abandoned well and, when solidified, the compositions is substantially immobilized within the formation. Compounds suitable as crosslinking agents are collected in Table 75.

To control the solidifying time of the drilling mud, a crosslinking reaction regulator is optionally also added to the drilling mud. Exemplary crosslinking reaction regulators include reducing agents that are capable of activating the crosslinking agent, sequestering agents that are capable of inhibiting the activity of the crosslinking agent and are releasing the

Table 75 Polymers and crosslinking agents for mud disposal [39]

Polymers	Crosslinking agents
Hydroxypropyl guar	Aldehydes
Polyacrylamides	Dialdehydes
Xanthan gum	Phenols
Poly(vinyl alcohol)	Ethers
Poly(vinyl acetate)	Aluminates
Poly(vinylpyrrolidone)	Gallates
Polyalkyleneoxides	Titanium chelates
Carboxycelluloses	Aluminum citrate
Hydroxyethyl cellulose	Chromium citrate
Galactomannans	Chromium acetate
Acrylic acid acrylamide copolymers	Chromium propionate

Table 76 Reducing agents for mud disposal [39]

Sulfur containing	Sulfur free
Sodium sulfite	Hydroquinone
Sodium hydrosulfite	Ferrous chloride
Sodium metabisulfite	p-Hydrazinobenzoic acid
Potassium metabisulfite	Hydrazine phosphite
Sodium sulfide	Hydrazine dichloride
Sodium thiosulfate	Manganese chloride
Ferrous sulfate thioacetamide	Potassium iodide
Hydrogen sulfide	Potassium ferrocyanide
	Manganese nitrate

crosslinking agent at the desired conditions, or pH modifiers that are degrading at the desired conditions to adjust the pH so that the crosslinking agent goes in action. Typical reducing agents are shown in Table 76.

Sequestering agents are citrate, propionate, and acetate salts of polyvalent metal ions such as aluminum, chromium, and iron. pH modifying agents, are acid precursors and base precursors, which generally either hydrolyze or thermally decompose to form an acid or a base, respectively.

Typical classes of acid precursors include hydrolyzable esters, acid anhydrides, sulfonates, organic halides, and salts of a strong acid and a weak base. Exemplary precursors are shown in Table 77.

Acetin is a synonym for glycerol monoacetate, as well as Diacetin is a synonym for glycerol diacetate. Some precursors are shown in Figure 65.

After the crosslinkable polymer, the crosslinking agent, and the ingredients are mixed with the drilling mud, the resulting solidifiable, disposable drilling mud composition is injected into a subterranean formation [39].

Table 77 Acid and base precursors [39]

Acid precursor	Base precursor
Ethyl formate	Ammonium salts
Propyl formate	Quaternary ammonium salts
Ethyl acetate	Ammonium chloride
Glycerol monoacetate	Urea
Acetin	Thiourea
Glycerol diacetate	Substituted ureas
Diacetin	
Xanthanes	
Thiocyanates	
Poly(ethylene) esters	
Ethyl acetate esters	
Acrylate copolymers	
Dimethyl esters	

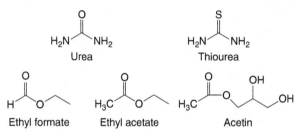

Figure 65 Acid and base precursors.

In detail, the composition is injected into the interior bore of the well, passes down the wellbore, and out at least one port in the well that is in fluid communication with the formation and the interior bore. After a some time, the composition solidifies and occupies a substantially fixed portion of the formation. Because the drilling mud is immobilized in the formation, adverse potential environmental risks due to the migration of the mud into other parts of the formation, such as potable water aquifers, is substantially reduced [39].

8.2 Recovery of Oil

The recovery of oil and from oil-in-water emulsion can be achieved by the use of biopolymers [40]. Biopolymers such as chitin and chitosan are low cost sorbents for the recovery of oil. Chitin has a greater adsorption capacity than chitosan due to its hydrophobic nature.

A systematic study was performed in the batch equilibrium mode to optimize various parameters, i.e., contact time, pH, dosage, initial concentration of oil, and temperature.

The adsorption process reached the equilibrium after 40 min of contact time and the removal of oil was found to be higher than 90% in an acidic medium.

The Freundlich and Langmuir isotherms were used to characterize the equilibrium. Therefrom, thermodynamic parameters such as ΔG, ΔH, and ΔS were calculated to elucidate the nature of the mechanism of sorption. In addition, kinetic studies were performed and tested with reaction-based and diffusion-based models.

9. DRILL CUTTING WASTES

The handling of drill cuttings and other wastes generated by drilling operations is both an environmental and economic issue [41]. The environmental regulations suggest to reinject the drill cuttings into the subsurface. This allows to achieve a zero discharge because oily cuttings are returned to their place of origin.

When this technology was originally introduced, an injection into a single well had a maximum slurry volume of some $5000\,m^3$. Now, particularly in very large projects, the amount of slurry to be injected has been grown up significantly.

Much progress has been made in the design of slurry rheology and the selection of the operational procedure, such as suspension and displacement to avoid a loss of injectivity, to maximize the capacity of disposal, and to minimize health, safety, and environment issues.

A risk-based approach that integrates deterministic software and tools, available data, knowledge, and experience, for the modeling of uncertainties and potential risks has been presented [41].

9.1 Microwave Treatment

Drilled cuttings contaminated by non aqueous fluids are a usual waste generated by oil well drilling activities [42]. More restrictive environmental legislation and the need to reduce drilling costs, both exploration and production, have led to the optimization of the solid–liquid separation. At the moment, in Brazil, the environmental offshore discharge threshold for cuttings contaminated by synthetic fluids is 6.9% by mass.

Microwave radiation is a promising technology to remove the organic phase of these drilled cuttings. A microwave heating technique has been presented as a method for the treatment of drilled cuttings contaminated by non aqueous fluids. This drying methodology was applied to contaminated cuttings and studied through a central composite design of experiments, where the variables were drilling fluid concentration, mass of cuttings, and specific energy. The responses were the residual paraffin and residual water contents [42].

9.2 Cuttings Transport

Drilling in massive salt zones is now a routine in the development of the sub-salt reservoirs offshore Brazil [43]. There, typical wells frequently include the drilling of 2000 m of salt zones with non aqueous fluids. Due to the regulations of discharge of cuttings, cutting drying processes must be done, which limit the drilling performance.

One alternative to enhance the efficiency of the process is to use saturated water-based fluids which would not require cuttings drying operations. The adoption of this technique requires special care with wellbore leaching and stability issues, but seems very attractive.

The design of hydraulic parameters to guarantee proper salt cuttings transport have been investigated. Since the cuttings are soluble in the drilling fluid, which is under saturated in downhole conditions, its diameter decrease with the flow, increasing the tolerance to cuttings concentration for a constant downhole pressure. As a consequence, the rate of penetration (ROP) limits are loosened and a safe hole cleaning can be achieved even at high ROPs [43].

9.3 Composting of Drill Cuttings

Drill cuttings that were contaminated with a non aqueous drilling fluid were co-composted with poultry manure and plant waste for 18 weeks [44]. A homogenized non aqueous-based fluid contaminated cutting was mixed with wood chips in a ratio of 1:1 and then mixed with soil, poultry, and plant waste manure in ratio of 4:2:1.

A heterotrophic bacterial population showed a steady increase up to 12 weeks. A mixed community population of the compost system was observed to be composed of 10 bacterial genera and 5 fungal genera. The highest recorded pH was 8.15. The electrical conductivity of the compost systems decreased continuously.

Based on these findings it has been recommended that oil exploration and production companies should adopt the compost technology with organic manure such as poultry and plant waste manure, as a waste management policy in order to reduce the high cost, energy, and pollution associated with other conventional treatment options [44].

10. MANAGEMENT OF PRODUCED WATER

As an oil field matures, it produces larger quantities of produced water [45]. Appropriate treatment levels and technologies depend on a number of factors, such as disposal methods or usage aims, environmental impacts, and economics. The key factors in the management of produced water and the goals emerge as [46]:

• Move toward zero emissions,
• No discharge to surface or seas,
• Waste-to-value conversion,
• Incremental and progressive separation, and
• Proactive efforts to influence partners, regulators, and environmental laws.

Technical approaches for the production, separation, and disposal of water injection and reservoir waterflooding procedures have been presented [46]. The best practices result both from both comprehensive assessments of the current available tools and from the insights obtained from previous experience.

A pilot plant with a capacity of $50\,m^3 day^{-1}$ was used to check the flotation, filtration, and adsorption trials for produced-water treatment at a crude oil gathering facility [45].

A flexible design of the plant allowed the testing of different combinations of these processes based on the special methods of treatment. Induced gas flotation trials were done with different rates of addition of the coagulant, poly(aluminum chloride). By coagulation/flocculation methods, dispersed oils were removed from the water.

The turbidity could be reduced by flotation, ranging from 57% to 78%. In addition, filtration further reduced the turbidity. Adsorption by activated carbon reduced the concentrations oil in water, as found by IR measurements.

The results indicated that an adsorption treatment would be more practical for water with a lower chemical oxygen demand, because high chemical oxygen concentrations in water would reduce the lifetime of the activated carbon dramatically [45].

11. OIL-POLLUTED GROUND

In the course of the transport of mineral oils, in particular crude oil, often grave pollution of the environment occurs due to oil spills and exiting oil [47]. Such events may have larger extents, for example, burst or leaking transport pipelines, but even smaller occurrences such as exiting fuel after accidents are problematic.

There are various tested methods for removing an oil contamination from the surface of water, for example, by an oil binding agent capable of floating, the cleaning of oil-contaminated soil is very difficult. A mechanical removal of the contaminated ground and its subsequent disposal is very costly and requires a lot of effort, even for very small amounts of contamination, and is financially and logistically impossible with greater amounts.

The cleaning of oil-contaminated surfaces is problematic, for example, on oil-contaminated rocky coasts and beaches, but also in the inside of oil tanks. A removal of the oil residues with water, e.g. with a high-pressure cleaner, may lead to a movement of the contamination being to a different location. A mechanical cleaning of surfaces on the other hand requires much effort [47].

An environmentally friendly treatment of oil-polluted stretches of ground and the cleaning of oil-contaminated surfaces can be done with a concentrate containing an emulsifier, a vegetable oil, and ethanol [47].

Non ionic tensides, in particular fatty amines, polyethylene glycols or oleic acid esters of ethoxylized castor oil are particularly suitable as emulsifiers. The composition of a concentrate for cleaning is shown in Table 78.

The concentrate from Table 78 may be used for manufacturing a pre-solution which is suitable for storage and transport. There, it is emulsified with water.

For its application as a treatment for stretches of ground, the concentrate is again diluted with water to form a cleaning solution and is applied to the oil-contaminated surface that is to be treated. The oil is emulsified in the cleaning solution and is detached from the particles of the ground.

Table 78 Composition of a concentrate [47]

Compound	Amount (vol.%)
Castor oil dioleate	12.5
Maize germ oil	12.5
Ethanol or propanol	75

Table 79 Tradenames in references

Tradename Description	Supplier
Marlowet® LVS PEG 18 castor oil dioleate [47]	SASOL Germany GmbH
Microtox® Standardized toxicity test system [7]	AZUR Environmental, MW Monitoring IP, Beckman Instruments, Inc.

The oil degrades in a significantly easier manner in the emulsified state. In particular, to clean oil-contaminated surfaces, the cleaning solution is sprayed onto the surfaces. Also, a powder-like adsorption agent can be added to the oil emulsion that is formed during the cleaning of oil-tank interiors. The adsorption agent adsorbs the emulsified oil and forms a precipitated sediment [47].

Tradenames appearing in the references are shown in Table 79.

REFERENCES

[1] Meinhold AF. Framework for a comparative environmental assessment of drilling fluids. Brookhaven Nat Lab Rep, BNL-66108; Brookhaven Nat Lab; 1998.

[2] CWA. Clean Water Act. Law CWA. Washington, DC: US Environmental Protection Agency; 2012. URL: http://water.epa.gov/action/cleanwater40/cwa101.cfm.

[3] SWDA. Safe Drinking Water Act. Law SWDA. Washington, DC: US Environmental Protection Agency; 2012. URL: http://water.epa.gov/lawsregs/rulesregs/sdwa/.

[4] Gallivan D. Hydraulic fracturing: the intersection of commerce, property rights and the environment in New York state. Law School Student Scholarship; 2014; Paper 470. URL: http://scholarship.shu.edu/student_scholarship/470.

[5] Dorn P, Rabke S, Glickman A, Nguyen K, MacGregor R, Candler J, et al. Development, verification, and improvement of a sediment toxicity test for regulatory compliance. SPE Drill Completion 2007;22(2). doi:10.2118/94269-PA.

[6] EPA Toxicity Test. Mysid acute toxicity test. Toxicity Test. EPA 712-C-96-136. Washington, DC: US Environmental Protection Agency; 1996. URL: http://www.epa.gov/ocspp/pubs/frs/publications/OPPTS_Harmonized/850_Ecological_Effects_Test_Guidelines/Drafts/850-1035.pdf.

[7] Patel AD. Low toxicity shale hydration inhibition agent and method of use. US Patent 8,298,996, assigned to M-I L.L.C. (Houston, TX); 2012. URL: http://www.freepatentsonline.com/8298996.html.

[8] Johnson BT. Microtox® acute toxicity test. In: Blaise C, Férard JF, editors. Small-scale freshwater toxicity investigations. Netherlands: Springer. ISBN 978-1-4020-3119-9; 2005, p. 69–105. doi:10.1007/1-4020-3120-3_2.

[9] AZUR Environmental. Microtox® acute toxicity test; 2010. URL: http://www.coastalbio.com/images/Acute_Overview.pdf.

[10] Bujdakova H, Kuchta T, Sidoova E, Gvozdjakova A. Anti-Candida activity of four antifungal benzothiazoles. FEMS Microbiol Lett 1993;112(3):329–34. URL: http://www.scopus.com/inward/record.url?eid=2-s2.0-0027327305&partnerID=40&md5=81aa10882351cb00e18bd6e65a05e2db.

[11] Chen CC, Lin CE. Analysis of copper corrosion inhibitors by capillary zone electrophoresis. Anal Chim Acta 1996;321(2-3):215–8. doi:10.1016/0003-2670(95)00591-9.

[12] Habibi MH, Tangestaninejad S, Yadollahi B. Photocatalytic mineralisation of mercaptans as environmental pollutants in aquatic system using TiO_2 suspension. Appl Catal B: Environ 2001;33(1):57–63. doi:10.1016/S0926-3373(01)00158-8.

[13] Fiehn O, Wegener G, Jochimsen J, Jekel M. Analysis of the ozonation of 2-mercaptobenzothiazole in water and tannery wastewater using sum parameters, liquid- and gas chromatography and capillary electrophoresis. Water Res 1998;32(4):1075–84. doi:10.1016/S0043-1354(97)00332-1.

[14] Hine CH, Kodama JK, Loquvam GS. The toxicity of allylamines. Arch Environ Health: Int J 1960;1(4):343–52. doi:10.1080/00039896.1960.10662707.

[15] Toraason M, Luken ME, Breitenstein M, Krueger JA, Biagini RE. Comparative toxicity of allylamine and acrolein in cultured myocytes and fibroblasts from neonatal rat heart. Toxicology 1989;56(1):107–17. doi:10.1016/0300-483X(89)90216-3.

[16] Duhon Sr JJ. Olive pulp additive in drilling operations. US Patent 5,801,127; 1998. URL: http://www.freepatentsonline.com/5801127.html.

[17] Ziemkiewicz P, Quaranta J, Darnell A, Wise R. Exposure pathways related to shale gas development and procedures for reducing environmental and public risk. J Nat Gas Sci Eng 2014;16:77–84. doi:10.1016/j.jngse.2013.11.003.

[18] St Clergy J, Toney FL. Methods for disposing of produced water recovered during hydrocarbon drilling, production or related operations. US Patent 8,608,405, assigned to Baker Hughes Incorporated (Houston, TX); 2013. URL: http://www.freepatentsonline.com/8608405.html.

[19] Wilson WN, Miles LH, Boyd BH, Carpenter RB. Cementing oil and gas wells using converted drilling fluid. US Patent 4,883,125, assigned to Atlantic Richfield Company (Los Angeles, CA); 1989. URL: http://www.freepatentsonline.com/4883125.html.

[20] Filippov L, Thomas F, Filippova I, Yvon J, Morillon-Jeanmaire A. Stabilization of NaCl-containing cuttings wastes in cement concrete by in situ formed mineral phases. J Hazard Mater 2009;171(1-3):731–8.

[21] Haque MN, Kayyali OA. Free and water soluble chloride in concrete. Cem Concr Res 1995;25(3):531–42.

[22] Al-Ansary MS, Al-Tabbaa A. Stabilisation/solidification of synthetic petroleum drill cuttings. J Hazard Mater 2007;141(2):410–21.

[23] Anonymous. Characterization of waste—leaching; compliance test for leaching of granular and sludges—part 1: One stage batch test at a liquid to solid ration of 2 l/kg with particle size below 4 mm (without or with size reduction). European Standard, EN 12457-1. Brussels: CEN—Committee for European Standardization; 2002.

[24] Wyant RE, Van Dyke O. Method and composition for cementing oil well casing. US Patent 3,499,491, assigned to Dresser Ind.; 1970. URL: http://www.freepatentsonline.com/3499491.html.

[25] Bell S. Mud-to-cement technology converts industry practices. Pet Eng Int 1993;65(9):51–2,54–5. URL: http://www.osti.gov/scitech/biblio/6121882.

[26] Cowan KM, Hale AH. High temperature well cementing with low grade blast furnace slag. US Patent 5,379,840, assigned to Shell Oil Company (Houston, TX); 1995. URL: http://www.freepatentsonline.com/5379840.html.

[27] Cowan KM, Hale AH, Nahm JJW. Dilution of drilling fluid in forming cement slurries. US Patent 5,314,022, assigned to Shell Oil Company (Houston, TX); 1994. URL: http://www.freepatentsonline.com/5314022.html.

[28]   Cowan KM, Smith TR. Application of drilling fluids to cement conversion with blast furnace slag in Canada. In: Proceedings volume, no. 93-601, CADE/CAODC Spring Drilling Conf. (Calgary, Can., 4/14–16/93) Proc.; 1993.

[29]   Zhao L, Xie Q, Luo Y, Sun Z, Xu S, Su H, et al. Utilization of slag mix mud conversion cement in the Karamay oilfield, Xinjiang. J Jianghan Pet Inst 1996;18(3): 63–6.

[30]   Caveny B, Ashford D, Hammack R, Garcia JG. Tires fuel oil field cement manufacturing. Oil Gas J 1998;96(35):64–7.

[31]   Schreiber Jr RJ, Yonley C. The use of spent catalyst as a raw material substitute in cement manufacturing. ACS Pet Chem Div Preprints 1993;38(1):97–9.

[32]   Hundebol S. Method and plant for manufacturing cement clinker. WO Patent 9429231; 1994.

[33]   Terry DT, Onan DD, Totten PL, King BJ. Converting drilling fluids to cementitious compositions. US Patent 5,295,543, assigned to Halliburton Company (Duncan, OK); 1994. URL: http://www.freepatentsonline.com/5295543.html.

[34]   Benge OG, Webster WW. Evaluation of blast furnace slag slurries for oilfield application. In: Proceedings volume. Iadc/SPE Drilling Conf. (Dallas, 2/15–18/94); 1994, p. 169–80.

[35]   Saasen A, Salmelid B, Blomberg N, Hansen K, Young SP, Justnes H. The use of blast furnace slag in North Sea cementing applications. In: Proceedings volume, vol. 1. SPE Europe Petrol. Conf. (London, UK, 10/25–27/94); 1994, p. 143–53.

[36]   McCarthy SM, Daulton DJ, Bosworth SJ. Blast furnace slag use reduces well completion cost. World Oil 1995;216(4):87–88,90,92,94,96.

[37]   Hale AH. Well drilling cuttings disposal. US Patent 5,341,882, assigned to Shell Oil Company (Houston, TX); 1994. URL: http://www.freepatentsonline.com/5341882. html.

[38]   Hofstätter H, Holzer C. Recycling of borehole solids in polymers. WO Patent 2013,037,978 assigned to Montanuniversität Leoben; 2013. URL: https://www. google.at/patents/WO2013037978A1?cl=en.

[39]   Dovan HT. Drilling mud disposal technique. US Patent 5,213,446, assigned to Union Oil Company of California (Los Angeles, CA); 1993. URL: http://www. freepatentsonline.com/5213446.html.

[40]   Elanchezhiyan SSD, Sivasurian N, Meenakshi S. Recovery of oil from oil-in-water emulsion using biopolymers by adsorptive method. Int J Biol Macromol 2014. doi:10.1016/j.ijbiomac.2014.07.002.

[41]   Guo Q, Geehan T, Ovalle A. Increased assurance of drill cuttings reinjection: challenges, recent advances, and case studies. SPE Drill Completion 2007;22(2). doi:10.2118/87972-PA.

[42]   Pereira MS, de Ávila Panisset CM, Martins AL, Marques de Sá CH, de Souza Barrozo MA, Ataíde CH. Microwave treatment of drilled cuttings contaminated by synthetic drilling fluid. Sep Purif Technol 2014;124(0):68–73. URL: http://www. sciencedirect.com/science/article/pii/S138358661400032X. doi:10.1016/j.seppur. 2014.01.011.

[43]   Silva F, Calçada L. Transport of soluble drilled cuttings. American Association of Drilling Engineers 2014; AADE-14-FTCE-45. URL: http://www.aade.org/app/ download/7238001550/AADE-14-FTCE-45.pdf.

[44]   Imarhiagbe EE, Atuanya EI, Ogiehor IS. Co-composting of non-aqueous drilling fluid contaminated cuttings from ologbo active oilfield with organic manure. Niger J Biotechnol 2013;26. URL: http://www.ajol.info/index.php/njb/article/ view/103388.

[45] Al-Maamari RS, Sueyoshi M, Tasaki M, Okamura K, Al-Lawati Y, Nabulsi R, et al. Flotation, filtration, and adsorption: pilot trials for oilfield produced-water treatment. Oil Gas Facilities 2014;2(3):56–66.

[46] Abou-Sayed A, Zaki K, Wang G, Sarfare M, Harris M. Produced water management strategy and water injection best practices: design, performance, and monitoring. SPE Prod Oper 2007;22(1). doi:10.2118/108238-PA.

[47] Kroh W. Agent for treating oil-polluted ground, and for cleaning oil-contaminated surfaces and containers. US Patent 7,947,641, assigned to Swisstech Holding AG (Zug, CH); 2011. URL: http://www.freepatentsonline.com/7947641.html.

INDEX

Note: Page numbers in **Bold** refer to figures.

Triton® X (Series)
  Poly(alkylene oxide), nonionic
    surfactants, 98, 245
Triton® X-100
  Hydrophilic poly(ethylene oxide), 245
Twaron®
  Aramid, 98
Tween® (Series)
  Ethoxylated fatty acid ester surfactants,
    166
Tween® 20
  Sorbitan monolaurate, 166
Tween® 21
  Sorbitan monolaurate, 166
Tween® 40
  Sorbitan monopalmitate, 166
Tween® 60
  Sorbitan monostearate, 166
Tween® 61
  Sorbitan monostearate, 166
Tween® 65
  Sorbitan tristearate, 166
Tween® 81
  Sorbitan monooleate, 166
Tween® 85
  Sorbitan monooleate, 166
Tychem® 68710
  Carboxylated styrene/butadiene
    copolymer, 98
Tylac® CPS 812
  Carboxylated styrene/butadiene
    copolymer, 98
Unamide®
  Polyoxyalkylated fatty amides, 98
Versatec™
  Oil-based mud, 98
Versatrol™ HT
  Gilsonite, 98
VES-STA 1
  Gel stabilizer, 166
Well life® (Series)
  Fibers for cement improvement, 167
Wellguard™ 7137
  Interhalogen gel breaker, 167
Westvaco® Diacid
  Diels-Alder acylating agents, 98

WG-3L VES-AROMOX® APA-T
  Viscoelastic surfactant, 167
WS-44
  Emulsifier, 167
XAN-PLEX™ D
  Polysaccharide viscosifying polymer, 98
Xanvis™
  Polysaccharide viscosifying polymer, 98
Ziboxan®
  Xanthan gum, 98

Acronyms
AA
  Acrylic acid, 34
AAm
  Acrylamide, 30
AMPS
  2-Acrylamido-2-methyl-1-propane
    sulfonic acid, 34
ANN
  Artificial neural network, 62
CMC
  Carboxymethyl cellulose, 37
CWA
  Clean Water Act, 251, 252
EDTA
  Ethylene diamine tetraacetic acid, 127
EG
  Ethylene glycol, 139
EO
  Ethylene oxide, 80
HEC
  Hydroxyethyl cellulose, 34
HPAN
  Hydrolyzed poly(acrylonitrile), 37
HTHP
  High temperature/high pressure, 31
MA
  Maleic anhydride, 138
  Methacrylic acid, 238
NVP
  N-Vinylpyridine, 31
OBM
  Oil-based drilling mud, 12
PAC
  Polyanionic cellulose, 34

Chemicals

Printed in the United States
By Bookmasters